Harmonic Analysis: A Gentle Introduction

Carl L. DeVito, PhD

University of Arizona

JONES AND BARTLETT PUBLISHERS

Sudbury, Massachusetts

BOSTON   TORONTO   LONDON   SINGAPORE

World Headquarters

Jones and Bartlett
 Publishers
40 Tall Pine Drive
Sudbury, MA 01776
978-443-5000
info@jbpub.com
www.jbpub.com

Jones and Bartlett Publishers
 Canada
6339 Ormindale Way
Mississauga, Ontario
L5V 1J2
CANADA

Jones and Bartlett Publishers
 International
Barb House, Barb Mews
London W6 7PA
UK

Jones and Bartlett's books and products are available through most bookstores and online booksellers. To contact Jones and Bartlett Publishers directly, call 800-832-0034, fax 978-443-8000, or visit our website www.jbpub.com.

> Substantial discounts on bulk quantities of Jones and Bartlett's publications are available to corporations, professional associations, and other qualified organizations. For details and specific discount information, contact the special sales department at Jones and Bartlett via the above contact information or send an email to specialsales@jbpub.com.

Copyright © 2007 by Jones and Bartlett Publishers, Inc.

ISBN-13: 978-0-7637-3893-8
ISBN-10: 0-7637-3893-X

All rights reserved. No part of the material protected by this copyright may be reproduced or utilized in any form, electronic or mechanical, including photocopying, recording, or by any information storage and retrieval system, without written permission from the copyright owner.

Library of Congress Cataloging-in-Publication Data

DeVito, Carl L.
  Harmonic analysis : a gentle introduction / Carl L. DeVito.
     p. cm.
  Includes bibliographical references and index.
  ISBN-13: 978-0-7637-3893-8
  ISBN-10: 0-7637-3893-X
  1. Harmonic analysis. I. Title.
  QA403.D457 2007
  515'.2433—dc22

                              2006010535

6048

Production Credits
Acquisitions Editor: Tim Anderson
Production Director: Amy Rose
Editorial Assistant: Laura Pagluica
Production Editor: Tracey Chapman
Production Assistant: Jamie Chase
Manufacturing Buyer: Therese Connell
Marketing Manager: Andrea DeFronzo
Composition: Northeast Compositors
Cover Design: Timothy Dziewit
Cover Image: © Bob Wainsworth/ShutterStock, Inc.
Printing and Binding: Malloy, Inc.
Cover Printing: John Pow Co.

Printed in the United States of America
10 09 08 07 06   10 9 8 7 6 5 4 3 2 1

Contents

Preface v

Chapter 0   **Preliminaries**   1
       1. Set Theory   1
       2. Relations and Functions   4
       3. The Real Number System   7
       4. The Complex Number System   11
       5. Analysis   15

Chapter 1   **Classical Harmonic Analysis**   23
       1. The Dirichlet Problem for a Disk   24
       2. Continuous Functions on the Unit Circle   25
       3. The Method of Fourier   29
       4. Uniform Convergence   32
       5. The Formulas of Euler   39
       6. Cesàro Convergence   43
       7. Fejér's Theorem   47
       8. At Last the Solution   52

Chapter 2   **Extensions of the Classical Theory**   57
       1. Functions on $(-\pi, \pi)$ 57
       2. Functions on Other Intervals   61
       3. Functions with Special Properties   65
       4. Pointwise Convergence of the Fourier Series   70

Chapter 3   **Fourier Series in Hilbert Space   79**
  1. Normed Vector Spaces   80
  2. Convergence in Normed Spaces   84
  3. Inner Product Spaces   90
  4. Infinite Orthonormal Sets, Hilbert Space   96
  5. The Completion   103
  6. Wavelets   107

Chapter 4   **The Fourier Transform   113**
  1. The Fourier Transform on Z   113
  2. Invertible Elements in $\ell^1(Z)$ 119
  3. The Fourier Transform on R   123
  4. Naïve Group Theory   128
  5. Not So Naïve Group Theory   131
  6. Finite Fourier Transform   135
  7. An Application   143
  8. Some Algebraic Matters   150
  9. Prime Numbers   154
  10. Euler's Phi Function   159

Chapter 5   **Abstract Algebra   165**
  1. Groups   165
  2. Morphisms   169
  3. Rings   172
  4. Fields   175

Appendix A   **Linear Algebra   179**

Appendix B   **The Completion   181**

Appendix C   **Solutions to the Starred Problems   187**

Index   213

Preface

Many branches of mathematics come together in harmonic analysis, each adding richness to the subject and each giving insights into this fascinating field. My purpose in writing this book is to give the reader an appreciation for all this interaction without overwhelming him or her with technical details. I also hope to inspire the reader to learn more about harmonic analysis and about mathematics in general.

This book can be read by any student who has had a course in advanced calculus and has some acquaintance with linear algebra. My students were mostly mathematics majors in their junior or senior year. Fourier transforms, the fast Fourier transform, even abstract algebra, have all made their way into modern engineering and into many areas of modern science. Some of my students were majoring in these subjects. For them, I usually asked that they knew some linear algebra and had previously taken an "Advanced Analysis for Engineers" type of course. Typically these students had trouble with the proofs, but, they told me, they had little interest in the proofs and were more concerned with understanding the definitions and doing the calculations. Because of the varied backgrounds of my students, I have tried to make the book as self-contained as possible.

The text begins with a very elementary preliminary chapter (Chapter Zero). This contains the kind of material that every student has seen but many have forgotten. It is there mainly for reference so that a student who has forgotten some fact about complex numbers, or some theorem from calculus, can refresh his or her memory. The only thing that might be new to some readers is the derivation of the Cauchy-Riemann equations. These are used to convince the student that Laplace's equation has lots of particular solutions. We refer to this fact in Chapter One where we define harmonic functions. It also tells the student why solving partial differential equations requires a new technique.

Chapter One contains a discussion of the Dirichlet problem for a circle with continuous boundary data. Here harmonic functions and periodic functions are treated. We also touch on ordinary differential equations and the method of separation of variables for partial differential equations. There is a section dealing with uniform convergence. This can, of course, be skipped or assigned as reading if the students are familiar with this topic. We use this material to derive the Euler formulas. The theorem of Fejér, that the Cesàro means of the Fourier series of a continuous function on the circle converge to that function uniformly over the circle, is proved in Section 7. We use that result to solve the Dirichlet problem and to prove the Weierstrass approximation theorem.

The purpose of Chapter One is to give the students a thorough foundation in basic analysis as it relates to trigonometric series. Once this has been mastered, the student is ready to learn something about the applications of these series and about the more abstract approaches to their study.

In Chapter Two we discuss the way a trigonometric series can be used to represent functions of arbitrary period, functions defined only on an interval, and functions with discontinuities. Lots of examples of functions and their Fourier series are given. Much of this material can be covered quickly, and it is not used in any subsequent section. There is, however, a theorem about pointwise convergence that will be new to most students.

In Chapter Three we look at Fourier series in Hilbert space. Here a little linear algebra is needed, and a brief discussion of the material required is given in Appendix A. We discuss inner product spaces and normed spaces and their interrelation. Banach and Hilbert spaces are defined and their basic properties investigated. A proof that every normed space has a completion that is a Banach space is given in Appendix B. There we also prove that the completion of an inner product space is a Hilbert space. These proofs are lengthy and a bit tedious, which is why we put them in an appendix. Instructors can cover them or simply assign them as reading.

In the remainder of Chapter Three, we define $L^2(T)$, T being the unit circle, to be the completion of the continuous functions on T, $C(T)$, with the integral inner product. We then prove that $\{e^{int}/\sqrt{2\pi}\}$ is an orthonormal basis for this Hilbert space. Once again the theorem of Fejér is used. We end Chapter Three with a brief discussion of wavelets. Here we define the Hilbert space $L^2(\mathbb{R})$ and show how certain functions give rise to orthonormal bases for this space.

Chapter Four deals with the Fourier transform. We start with the vector space $\ell^1(Z)$; sequences $\{a(n)\}$ such that $\sum |a(n)| < \infty$. We define the convolution operation and the Fourier transform for elements of this space. The advantage in doing that is that all the proofs can be given, and in this case they are quite transparent.

Next we look at elements of $\ell^1(Z)$ that are invertible with respect to convolution. To show that there is a connection between these ideas and Fourier series, we discuss Wiener's theorem: If $f \in C(T)$ is never zero and the Fourier series of f is absolutely convergent, then the Fourier series of $1/f$ is absolutely convergent. This connection motivates a further investigation into the properties of $\ell^1(Z)$ under convolution, and we end Section 2 with a sketch of the proof of Wiener's result. Although the proof is, necessarily, incomplete, I believe I have done enough to give the reader a pretty good idea of what is involved.

In Section 3 we discuss the Fourier transform on $\mathbb{R}$. Because we are *not* assuming a knowledge of the Lebesgue integral we pursue, as far as possible, the analogy with $\ell^1(Z)$. Many examples of functions and their Fourier transforms are presented, and we show, by means of an example, how the Fourier transform can be used to solve a differential equation.

Reviewing the first three sections of this chapter we observe that the Fourier transform involves maps from Z into T and continuous maps from $\mathbb{R}$ into T, that "preserve the algebra" [i.e., maps c such that $c(m+n) = c(m) \cdot c(n)$]. This, of course, takes us into group theory and the notion of a homomorphism. We do not use this terminology, however, until Chapter Five. Here we define, very simply, subsets of $\mathbb{C}$ that are groups under addition or under multiplication. The discussion soon leads us to Z_n (the integers modulo n) and its additive structure. By investigating maps from Z_n into T that "preserve the algebra" we are led to the group of nth roots of unity and the finite Fourier transform. This transform is of central importance in certain branches of engineering (e.g., digital signal processing). We discuss its properties and show how it arises, rather mysteriously, in Lagrange's treatment of equations. This is done in Section 7. The calculations are a little lengthy, but this material is not used in later sections and so it can be skipped or assigned as reading.

Sections 8–10 contain some number theory. In Section 8, we show when a power of a primitive nth root of unity is again primitive, we discuss the multiplicative structure of Z_n and characterize those elements that have inverses, and we define the Euler phi function. Some instructors may want to skip sections 9 and 10 and go directly to Chapter Five. Finding a formula for the phi function requires us to discuss primes, and to keep things interesting, we mention perfect numbers and their relation to Mersenne primes. At this point we have done enough modular arithmetic to prove the interesting half of the Lucas-Lehmer test, which tells us when $2^p - 1$ is a prime. We end Chapter Four with a brief discussion of how the phi function enters into the creation of cyphers.

The final chapter of the book contains the basic definitions from abstract algebra. Here we define groups and point out the many examples of groups that have occurred at various places in the text. We define homomorphisms and isomorphisms and explain that the maps from Z into T, discussed in connection with the Fourier transform, are examples of these things. The Fourier transform itself is an isomorphism between rings, and so we define rings and fields and point out these facts. A short treatment of the properties of finite fields (these do arise in the applications of harmonic analysis) is also given.

The detailed discussion of the Dirichlet problem given in Chapter One is intended as motivation for the study of trigonometric series. Instructors who wish to by-pass this material and begin their course with the series themselves may easily do so. Simply cover Sections 4, 5, 6, and 7 of Chapter One and omit Section 8. The remainder of the text is then accessible.

Those instructors who want to get to the Fourier Transform as rapidly as possible may do so by covering Sections 4–7 of Chapter One, then Sections 1 and 2 of Chapter Three, and the material in Section 5 up to Theorem 4 (in particular, Definition 1 and Theorem 3). I have tried to get the reader actively involved in the development of the mathematics. To accomplish this, I put some results, results referred to later in the text, in the exercises. Any such exercise is marked with a star and should be worked, or at least read, by anyone using the book. Solutions to the starred problems are given in Appendix C.

I received helpful suggestions about various parts of the book from a great many people. I would now like to thank Nirmal Bose (Pennsylvania State University), James Kang (California State Polytechnic University), David Walnut (George Mason University), Selin Aviyente (Michigan State University), Kasso Okoudjou (Cornell University), Ilya Pollak (Purdue University), and John Brillhart and William Farris (University of Arlizona). It is my pleasure to also thank my Belgian colleague Maurice Hasson for reading the entire manuscript and making many valuable suggestions. I would also like to thank Victoria Milne for typing the manuscript and having infinite patience during its many revisions.

Chapter Zero

Preliminaries

All the mathematics needed to read this book is outlined in this chapter. The advantage in doing this is that the material is then easy for us to reference and, if necessary, easy for the reader to review. The notations and conventions used in the book are also explained here, but these are, for the most part, quite standard.

1. Set Theory

In ordinary English the word we use for a collection of objects depends on what those objects are. So a collection of cows is a herd, while a collection of sheep is a flock. A collection of owls is a parliament, while a collection of ravens is a murder. We even have a name for a collection of unicorns. It's called a blessing. In the terminology of mathematics the word "set" is used for any collection of objects, whatever those objects may be. These collections, however, must be well defined; i.e., it must be clear just which objects are in a given collection and which objects are not. The collection of all great novels, for example, is not a set because there is no general agreement, even among experts, as to what constitutes a "great" novel. Moreover, are we to include in our collection novels yet to be written or only those in print today? One faces similar difficulties with any collection of physical entities. The only "real" sets are collections of mathematical objects. Still, it is often illuminating to consider collections of physical entities and to call them sets provided one understands that this practice is really just a pedagogical device whose value is more psychological than mathematical.

The objects comprising a set are called its "members" or its "elements." Given a set X and an object x we write $x \in X$ (read "x belongs to X") when

x is a member of the set X, and we write $x \notin X$ when this is not the case. A set is completely determined by its elements. More formally:

Definition 1. Two sets X and Y are said to be equal, written $X = Y$, if every element of X is an element of Y and every element of Y is an element of X.

Sets that contain only a few objects can be specified by simply listing their members. To emphasize the fact that we want to consider the set, a new "entity" formed by the act of collecting these objects together, we list them between a pair of curly brackets. So $\{a, b, c\}$ is the set consisting of the three letters a, b, and c from the English alphabet. This notation is impractical when we work with sets that contain a great many members. The next definition will help us deal with that situation.

Definition 2. Let X, Y be two sets. We shall say that X is a subset of Y, and write $X \subseteq Y$, if every element of X is also an element of Y.

When $X \subseteq Y$, we shall say that X is contained in or included in Y, and we shall say that Y contains, or includes, X. It is clear that $X = Y$ if, and only if, we have both $X \subseteq Y$ and $Y \subseteq X$. This observation is useful because in many problems the easiest way to show that two sets are equal is to show that each contains the other. (See Exercises 2, problem 1b.)

It is often convenient to define a set by specifying that its members are all elements of a known set, X say, and that they are distinguished from the other elements of X by the fact that they satisfy a certain condition. For example, if X is the set of all atoms, we may want to consider the subset of X defined as follows: $\{x \in X | x \text{ is a hydrogen atom}\}$. The vertical bar is read "such that" and the total expression reads "the set of all x in X such that x is a hydrogen atom." When this notation is used, it is assumed that the condition specified is clear enough to define a unique subcollection of X.

Definition 3. Given two sets X and Y we define two new sets as follows: (a) $X \cup Y$ (read "X union Y") is the set consisting of those objects that are members of X or of Y, and (b) $X \cap Y$ (read "X intersect Y") is the set consisting of those objects that are members of X and Y.

It is clear that $X \cup Y$ contains both X and Y and that $X \cap Y$ is contained in both X and Y. Moreover, because we use "or" in the inclusive sense, we always have $X \cap Y \subseteq X \cup Y$. The set with no members is called the empty set and is denoted by $\emptyset$. From Definition 2 we see that $\emptyset$ is a subset of any set X. This is consistent with Definition 3 because $\emptyset = \emptyset \cap X \subseteq X$.

Definition 4. Two sets X and Y are said to be disjoint if $X \cap Y = \emptyset$. It is a curious fact that the empty set is contained in, yet disjoint from, every set.

A set that contains a single element, say x, is called a singleton, or the singleton defined by x. We denote this set by $\{x\}$. If y is another object different from x, then the set $\{x, y\}$ is called the unordered pair defined by x and y. This terminology reflects the fact that $\{x, y\} = \{y, x\}$. We shall need to consider the ordered pair defined by x and y.

Definition 5. Given two objects x and y we define their ordered pair, (x, y), to be the set $\{\{x\}, \{x, y\}\}$. Given two sets X and Y, the set $X \times Y = \{(x, y) |\, x \in X$ and $y \in Y\}$ is called the Cartesian product of X and Y.

We leave it to the reader to show that $(x, y) \neq (y, x)$ unless we happen to have $x = y$ (Exercises 1, problem 2). The Cartesian product of a set X with itself, $X \times X$, is sometimes denoted by X^2.

Exercises 1

A number of useful facts, facts that are referred to later in the text, are scattered among the exercises. In some of the exercises we introduce terminology that is also used later. Any problem that is referred to later in the text is marked with an asterisk.

1. Show that $\varnothing$ is not equal to $\{\varnothing\}$ and that neither of these sets is equal to $\{\varnothing, \{\varnothing\}\}$.

*2. Given two objects x and y we have defined (x, y) to be the set $\{\{x\}, \{x, y\}\}$.

    (a) Show that $(x, y) = (y, x)$ if, and only if, $x = y$.
    (b) Show that $(x, y) = (u, v)$ if, and only if, $x = u$ and $y = v$.

    This exercise justifies our calling (x, y) the ordered pair defined by x and y.

*3. Let X, Y be two sets. We define $X - Y$ to be $\{x \in X | x \notin Y\}$. Show that $X - Y = X - (X \cap Y)$.

*4. Let E be a fixed set and let X be any subset of E. The set $E - X$ is called the complement of X (in E) and is denoted $C_E(X)$. Let X, Y denote arbitrary subsets of E.

    (a) Prove DeMorgan's theorem; i.e., show that
        i. $C_E(X \cup Y) = C_E(X) \cap C_E(Y)$;
        ii. $C_E(X \cap Y) = C_E(X) \cup C_E(Y)$.
    (b) Show that $X - Y = X \cap [C_E(X \cap Y)] = X \cap C_E(Y)$.
    (c) Show that the following are equivalent:
        i. $X \cup Y = E$
        ii. $C_E(X) \subseteq Y$
        iii. $C_E(Y) \subseteq X$

(d) Show that the following are equivalent:
   i. $X \subseteq Y$
   ii. $C_E(X) \supseteq C_E(Y)$
   iii. $X \cup Y = Y$
   iv. $X \cap Y = X$

2. Relations and Functions

We want to introduce some concepts, and the terminology associated with them, that will be useful throughout the book. To enhance the applicability of these ideas, we present them in a very general abstract setting. Some concrete examples, however, are given here and many more are given in subsequent sections.

Throughout this section, X and Y denote nonempty sets and $X \times Y$ denotes their Cartesian product (Section 1, Definition 5).

Definition 1. Any nonempty subset R of $X \times Y$ is called a relation from X to Y. We shall say that $x \in X$ and $y \in Y$ are R related, and we shall write xRy, if $(x, y) \in R$. The two sets $\mathcal{D}(R) \subseteq X$ and $\mathcal{R}(R) \subseteq Y$, called, respectively, the domain and the range of R, are defined as follows: $\mathcal{D}(R) = \{x \in X |$ there is a $y \in Y$ with $(x, y) \in R\}$, $\mathcal{R}(R) = \{y \in Y |$ there is an $x \in X$ with $(x, y) \in R\}$.

We can define a relation from the set M of all men to the set W of all women by taking $R = \{(m, w) \in M \times W |\ m$ and w are married (to each other)$\}$. Then $\mathcal{D}(R)$ is the set of all married men and $\mathcal{R}(R)$ is the set of all married women.

A relation from a set X to itself (i.e., a subset of $X \times X$) is called a relation on X. A relation R on X is said to be (a) reflexive, if xRx for every $x \in X$; (b) symmetric, if x, y in X and xRy implies yRx; (c) transitive, if for x, y, z in X, we have xRz whenever we have both xRy and yRz.

A relation that has all three of these properties (i.e., a relation that is reflexive, symmetric, and transitive) is called an equivalence relation on X.

Referring to the set W we can take $w \equiv w'$ to mean that the two women w and w' are sisters. (Let's agree that two, not necessarily different, women are sisters if they have the same parents.) It is easy to see that $\equiv$ is an equivalence relation on W.

Definition 2. A relation R from a set X to a set Y is called a function from X to Y or a mapping from X into Y, if (a) the domain of R is X and (b) whenever (x, y) and (x, y') are both in R, we must have $y = y'$.

We can rephrase (b) as xRy and xRy' together imply $y = y'$. A relation R that has property (b) but not (a) is sometimes called a partial function from X to Y. In this case, R is a function from X' to Y where $X' = \mathcal{D}(R)$.

In a monogamous, or a polyandrous, society, the relation from M to W defined above is a partial function. In a polygamous society, it is not. When f is a partial function from X to Y, it is customary to write $y = f(x)$ instead of xfy or $(x, y) \in f$. We shall also use the symbol $f : X \to Y$ to abbreviate the phrase "f is a function from X into Y." Let us emphasize that when we write $f : X \to Y$ the domain of f is all of X. The set Y, in this case, is often called the codomain of f and the range of f is, of course, a subset of Y.

Given $f : X \to Y$ we can, for each $y \in Y$, define a subset $\text{pre}_f(y)$ of X as follows: $\text{pre}_f(y) = \{x \in X|\, f(x) = y\}$. We call this set the preimage of y. Note that because f is a function, the sets $\text{pre}_f(y)$ and $\text{pre}_f(y')$ are disjoint unless $y = y'$. Moreover, $\text{pre}_f(y) \neq \emptyset$ if, and only if, y is in the range of f. There are two important special cases here. (i) When $f(x) = f(x')$ implies $x = x'$. In this case, every nonempty $\text{pre}_f(y)$ is a singleton. (ii) When the range of f is all of Y. In this case, no $\text{pre}_f(y)$ is empty.

Definition 3. A function f from X to Y is said to be one-to-one if $f(x) = f(x')$ implies $x = x'$. We shall say that f is onto or that f maps X onto Y if the range of f is Y.

Observe that when $f : X \to Y$ is both one-to-one and onto we can define a function $g : Y \to X$ by setting $g(y) = \text{pre}_f(y)$ for every $y \in Y$.

A set whose elements are themselves sets is often called a family of sets. Given a family of sets $\mathcal{A}$ we can always "name" the members of $\mathcal{A}$ as follows: Choose a set I and a one-to-one, onto function $\varphi : I \to \mathcal{A}$. We call φ an indexing of $\mathcal{A}$ and we call I the index set. It is customary to denote the set $\varphi(\lambda) \in \mathcal{A}$ by A_λ and to write $\mathcal{A} = \{A_\lambda|\, \lambda \in I\}$. We now define

$$\bigcup \mathcal{A} = \bigcup_{\lambda \in I} A_\lambda = \{x|\, x \in A_\lambda \text{ for some } \lambda \in I\}$$

and

$$\bigcap \mathcal{A} = \bigcap_{\lambda \in I} A_\lambda = \{x|\, x \in A_\lambda \text{ for every } \lambda \in I\}$$

A family of sets $\mathcal{A} = \{A_\lambda|\, \lambda \in I\}$ is said to be pairwise disjoint if $A_\lambda \cap A_\mu = \emptyset$ whenever λ, μ are in I and $\lambda \neq \mu$.

Given a function f from X to Y, the set $R(f)$ indexes the family of nonempty preimages of f; i.e., $\{\text{pre}_f(y)|\, y \in R(f)\}$. This is a family of nonempty pairwise disjoint sets whose union is X.

Axiom of Choice: When $f : X \to Y$ is onto it seems evident that there "should be" a function $g : Y \to X$ that is one-to-one; for each of the sets $\text{pre}_f(y)$ is nonempty so we may set $g(y)$ equal to any member of this set. This seems to give us a function, and because these preimages are pairwise disjoint, g must be one-to-one. The trouble with this is that we have not specified which element of $\mathrm{pre}_f(y)$ is to be $g(y)$; hence, we have not really defined anything here. There is, in fact, no way to specify $g(y)$ in the general case. To define g, we must appeal to a famous axiom: *Given any nonempty family of nonempty sets, there is a function that assigns to each set in the family of an element of that set.* This is the axiom of choice. It has many applications and many equivalent forms and we discuss some of them later (see Section 2 of Chapter 4).

Exercises 2

*1. Let $\equiv$ be an equivalence relation on a set X. For each $x \in X$ let $[x] = \{y \in X \,|\, x \equiv y\}$. We call $[x]$ the equivalence class defined by x.

   (a) Show that for each $x \in X$, $[x] \neq \emptyset$.
   (b) Show that for x, y in X, $[x] \cap [y] \neq \emptyset$ if, and only if, $[x] = [y]$.

   Thus the equivalence classes of X are either identical or they are disjoint.

*2. Let $f : X \to Y$ and let g be a partial function from Y to a set Z.

   (a) If the domain of g contains the range of f, show that setting $(g \circ f)(x) = g[f(x)]$ for each $x \in X$ defines a function, $g \circ f$, from X to Z. We call this the composition of f and g.
   (b) Show that f is both one-to-one and onto if, and only if, there is a function $g : Y \to X$ such that $(f \circ g)(y) = y$ for all $y \in Y$ and $(g \circ f)(x) = x$ for each $x \in X$. When such a g exists, we call it the inverse of f and denote it by f^{-1}.

*3. For any set X, let $P(X)$ denote the family of all subsets of X.

   (a) Define a relation $\equiv$ on $P(X)$ as follows: For $Y, Z \in P(X)$, $Y \equiv Z$ if, and only if, there is a function f from Z to Y that is both one-to-one and onto. Show that $\equiv$ is an equivalence relation on $P(X)$.
   (b) Define a relation $\leq$ on $P(X)$ as follows: For Y, Z in $P(X)$, $Y \leq Z$ if, and only if, $Y \subseteq Z$.
      i. Show that the relation $\leq$ is reflexive and transitive.
      ii. Show that $\leq$ is antisymmetric, meaning $Y \leq Z$ and $Z \leq Y$ together imply $Y = Z$.

      A relation R on a set X that is reflexive, antisymmetric, and transitive is called a partial order relation on X or a partial ordering of X. The term "partial" refers to the fact that there may be elements of X that are incomparable, i.e., elements x, y of X such that neither xRy nor yRx is true. A partial order relation on a set X is called a total order relation or a total ordering of X if any two elements of X are comparable.

4. Suppose that for some set X, we have a function $g : X \to P(X)$ that is onto. Then for each $x \in X$, $g(x) \in P(X)$ (i.e., $g(x)$ is a subset of X) and so it makes sense to ask whether or not x is an element of $g(x)$. Now let $X_0 = \{x \in X \mid x \notin g(x)\}$. This is a subset of X, hence an element of $P(X)$. There is an $x_0 \in X$ such that $g(x_0) = X_0$ because g is onto.

   (a) Show that $x_0 \in g(x_0)$ leads to a contradiction.

   (b) Show that $x_0 \notin g(x_0)$ leads to a contradiction.

   We conclude that for any set X there is no function from X to $P(X)$ that is onto.

   (c) The famous Russell paradox arises when we consider the "set" of all sets that are not members of themselves (e.g., the set of all books is not a book). If there were such a set, let's call it S, then we have both $S \in S$ and $S \notin S$. Paradoxes like this led to the creation of axiomatic set theory.

*5. Let $\{A_\lambda \mid \lambda \in I\}$ be a family of sets and suppose that there is a set X such that $A_\lambda \subseteq X$ for every $\lambda \in I$. Prove the strong form of DeMorgan's theorems, i.e., show that

   (a) $C_X \left[\bigcup_{\lambda \in I} A_\lambda \right] = \bigcap_{\lambda \in I} C_X(A_\lambda)$ and

   (b) $C_X \left[\bigcap_{\lambda \in I} A_\lambda \right] = \bigcup_{\lambda \in I} C_X(A_\lambda)$

3. The Real Number System

The real numbers were introduced into mathematics to provide a foundation for analysis (i.e., calculus and its off-shoots). We assume the reader has some familiarity with this set, but a brief review of its basic properties is well worthwhile.

To begin with, any two real numbers can be added and multiplied and the result is a real number. Moreover, this set, which we shall denote by $\mathbb{R}$, contains two special members 0 and 1 that have the properties:

(1) $x + 0 = 0 + x = x$ for all $x \in \mathbb{R}$

(2) $x \cdot 1 = 1 \cdot x = x$ for all $x \in \mathbb{R}$

These numbers, which are clearly unique (see Exercises 3, problem 1), are called, respectively, the additive and multiplicative identities of $\mathbb{R}$. Each element $x \in \mathbb{R}$ has a unique additive inverse, that is, there is a unique element $(-x) \in \mathbb{R}$ such that

(3) $x + (-x) = (-x) + x = 0$

Thus we may, for any two real numbers x and y, define $x - y$ to mean $x + (-y)$.

Any nonzero real number x has a unique multiplicative inverse. This is the number, we shall denote it by x^{-1}, for which

(4) $x \cdot x^{-1} = x^{-1} \cdot x = 1$

For any x, y in $\mathbb{R}$ with $y \neq 0$, we define x/y to mean $x \cdot y^{-1}$. We assume that the reader is familiar with the various properties of these arithmetic operations.

There is a subset P of $\mathbb{R}$ that has the following properties:

(5) For any $x \in \mathbb{R}$ exactly one of the statements $x \in P$, $(-x) \in P$, $x = 0$, is true.

(6) If x, y are in P so also are $x + y$ and $x \cdot y$.

It is easy to see that the number 1 is an element of the set P, whereas the number 0 is not. The members of P are called the positive real numbers. Given x, y, in $\mathbb{R}$ we say that x is less than y or y is greater than x, and write $x < y$, if $y - x$ belongs to P. This gives us a relation on the set $\mathbb{R}$ that we shall call the order relation on $\mathbb{R}$. It is clear from (5) that for any two distinct real numbers x and y we have $x < y$ or $y < x$ (but not both), and from (6) we see that $x < y$ implies $x + z < y + z$ for all $z \in \mathbb{R}$, and $x \cdot z < y \cdot z$ for all $z \in P$.

Because $1 \in P$ we see from (6) that $2 = 1 + 1$ is in P and so is $3 = 2 + 1$, etc. The subset of $\mathbb{R}$ generated in this way is denoted by $\mathbb{N}$ and is called the set of natural numbers. It has many fascinating properties but the one that we shall need in the sequel is this:

(7) Any nonempty set of natural numbers contains a smallest member.

The set $\mathbb{R}$ does not have property (7) because it has no smallest member. However, $\mathbb{R}$ does have an important property that can be stated in terms of the order relation. First, we need some terminology. Let S be a nonempty set of real numbers. A real number a such that $a \leq s$ (read "a is less than or equal to s") for all $s \in S$ is said to be a lower bound for S. Similarly, a number $b \in \mathbb{R}$ is said to be an upper bound for S if $s \leq b$ for all $s \in S$. A nonempty subset of $\mathbb{R}$ is said to be a bounded set if it has both an upper and a lower bound. By convention the empty set, $\varnothing$, is also called a bounded set.

Definition 1. Let S be a nonempty set of real numbers. A real number b_0 is said to be the (see Exercises 3, problem 2c) least upper bound for S if: (i) b_0 is an upper bound for S, and (ii) if b is any upper bound for S, then $b_0 \leq b$.

We may now state one of the key properties of the set $\mathbb{R}$.

(8) Any nonempty set of real numbers that has an upper bound has a least upper bound.

There are many ways to study the real numbers. In some treatments (8) is taken as an axiom and referred to as the axiom of continuity. In others it is a theorem and called the completeness property of $\mathbb{R}$. Let us show how some of the most familiar facts about $\mathbb{R}$ follow easily from (8). The results we have in mind are well known but, as we shall see in the sequel, extremely useful.

Theorem 1. Let a, b be any two positive real numbers. Then there is a natural number n such that $a < n \cdot b$.

Proof. If $a < b$, we need only take $n = 1$. Suppose that $nb \leq a$ for every natural number n. Then the set $\{nb|n \in \mathbb{N}\}$, which is certainly not empty, has an upper bound, the number a, and hence, by (8), has a least upper bound that we shall denote by a_0. Now for any $n \in \mathbb{N}$ we have $(n+1)b \leq a_0$, which implies the inequality $nb \leq a_0 - b$ for all $n \in \mathbb{N}$. Thus the number $a_0 - b$ is also an upper bound for the set $\{nb|n \in \mathbb{N}\}$, and this contradicts the fact that a_0 is the least upper bound for this set.

Corollary 1. Let a, b be any two positive real numbers. Then $a = n \cdot b + r$ where n is a natural number or zero, and r is a real number that satisfies the inequalities $0 \leq r < b$.

Proof. Suppose $b < a$. Then $\{n \in \mathbb{N}| a \leq nb\}$ is, by our theorem, nonempty and because it is a subset of $\mathbb{N}$, it contains a smallest member that we shall call m; here we are using property (7). Thus we may write $(m-1)b < a \leq mb$ and because $b < a$ we know that $2 \leq m$. It follows that the number $m - 1$, that we shall denote by n, is a natural number. Thus, except when $a = mb$ in which case $r = 0$, we have $a = (m-1)b + r = nb + r$ where r is the real number $a - nb$. Clearly, $r < b$ in either case.

Theorem 1 has another useful corollary. Before we can state it we must recall two important subsets of $\mathbb{R}$. First, we have the set of integers, Z, obtained from $\mathbb{N}$ by adjoining zero and the additive inverse of each of its members. So

(9) $Z = \{0, \pm 1, \pm 2, \dots\}$

Next there is the set Q of rational numbers,

(10) $Q = \{x \in R | x = m/n \text{ where } m, n \text{ are in } Z \text{ and } n \neq 0\}$

Corollary 2. Between any two distinct real numbers there is a rational number.

Proof. Let a, b be two real numbers with $a < b$. We want to find two integers m, n such that $a < \frac{m}{n} < b$.

Now the numbers 1 and $b - a$ are positive, so there is, by the theorem, a natural number n such that $1 < n(b - a)$. Choose, and fix, such an n. Now, again by our theorem, the set $\{k \in \mathbb{Z} | na < k\}$ is nonempty. Let m be its smallest member. Then $na < m \leq na + 1 < na + n(b - a) = nb$, giving us $a < \frac{m}{n} < b$ and proving the corollary.

The set $\mathbb{R}$ together with its order relation is, as we have just seen, a sophisticated mathematical object with an intricate structure. The members of $\mathbb{R}$ are lined up (linearly ordered) by this relation but not at all like the beads on a string. Between any two distinct real numbers there are always more real numbers. Thus given a specific number it is meaningless to speak of the "next" one. Moreover, removing a number does separate the line into two parts; it does leave a "gap," but this gap has no edges.

Exercises 3

1. (a) Show that the additive identity of $\mathbb{R}$ is unique; i.e., if o and o' both satisfy equation (1), show that $o = o'$.
   (b) Show that the multiplicative identity of $\mathbb{R}$ is unique.

2. (a) Given real numbers x and y with $x < y$, show that
      i. $x + z < y + z$ for any $z \in \mathbb{R}$, and
      ii. $x \cdot z < y \cdot z$ for any $z \in P$.
   (b) For x, y in $\mathbb{R}$ we have defined $x \leq y$ to mean $x < y$ or $x = y$. Show that $\leq$ is a total ordering of $\mathbb{R}$ (Exercise 2, problem 3b).
   (c) Show that a set of real numbers that has a least upper bound has exactly one least upper bound.
   (d) Show that any nonempty set $S \subseteq \mathbb{R}$ that has a lower bound has a greatest lower bound; i.e., there is a number a_0 such that
      i. a_0 is a lower bound for S, and
      ii. if a is any lower bound for S then $a \leq a_0$.
   (e) Leibnitz wanted to add infinitesimals to the real number system; i.e., quantities j such that $0 < j < \frac{1}{n}$ for *every* $n \in \mathbb{N}$. Show that no member of $\mathbb{R}$ has this property.

3. For any $x \in \mathbb{R}$ define $|x|$ to be x when $x \in P$ or $x = 0$ and define $|x|$ to be $(-x)$ otherwise. Show that for any x, y in $\mathbb{R}$ we have
   (a) $|x + y| \leq |x| + |y|$
   (b) $||x| - |y|| \leq |x - y|$
   (c) $|x \cdot y| = |x| \cdot |y|$

4. Suppose that for some set S we have a function $f : \mathbb{N} \to S$ that is onto. Show that there is a function $g : S \to \mathbb{N}$ that is one-to-one. Explain why you do *not* need the axiom of choice in this case.

5. (a) Show that there are real numbers that are not rational as follows: Consider the real number $\sqrt{2}$. If this were rational, there would be integers m, n such that $\sqrt{2} = \frac{m}{n}$. Clearly, we may suppose that m and n have no common factors. Now deduce a contradiction.

   (b) Let α be a real number that is not rational; such numbers are called irrational. Show that

      i. $\alpha + q$ is irrational for every $q \in Q$, and $\alpha \cdot q$ is irrational for every nonzero element of Q.

   (c) Given a, b in $\mathbb{R}$ with $a < b$, modify the proof of Corollary 2 to show that there are integers m, n for which $a < \left(\frac{m}{n}\right)\sqrt{2} < b$. Thus, between any two distinct real numbers, there is an irrational number.

6. For any two real numbers a and b, let $(a, b) = \{x \in \mathbb{R}|\, a < x < b\}$ and let $[a, b] = \{x \in \mathbb{R}|\, a \leq x \leq b\}$.

   (a) Under what conditions on a and b is the set (a, b) empty? What about the set $[a, b]$?

   (b) When $(a, b) \neq \varnothing$, we call this set an open interval, and when $[a, b] \neq \varnothing$ we call this set a closed interval. In either case, a and b are the end points of the interval and $b - a$ is its length. We have seen that every open interval contains both rational and irrational numbers. Is this true of every closed interval?

7. A purely fictitious sleep researcher discovers a man with a strange pattern of dreaming. On any closed interval of time during which he is asleep he dreams a great deal, but never on the (open) middle one-third of this interval. Suppose this man sleeps for 1 hour. Describe, as precisely as possible, the set on which his dreaming takes place. Hint: Model his nap by the interval $[0, 1]$. No dreaming takes place on the interval $(\frac{1}{3}, \frac{2}{3})$.

4. The Complex Number System

The reawakening of Europe, which began in the Italian city states, is usually associated with the remarkable achievements in painting, sculpture, and architecture. But it was an intellectual rebirth and so, not surprisingly, it had a mathematical side. Perhaps the two most important developments were the discovery of the laws of perspective by Brunelleschi, which led to the creation of projective geometry, and the discovery of the system of complex numbers. These numbers are of fundamental importance in modern physics and modern electrical engineering, but when they were first introduced they were viewed with skepticism and distrust. Let us briefly review how they came into being and give a quick survey of their basic properties.

One might believe that the study of the quadratic equation would lead to complex quantities. This, however, was not the case. Quadratics whose solution involved square roots of negative numbers were simply dismissed as having no solutions. It was in connection with the cubic equation that this could not be done. A formula for solving the general cubic was found by Scipione del Ferro and, independently, by Niccolo Fontana, also known as Ta'taglia. In 1545 this formula was published by Gerolomo Cardano in a book entitled *Ars Magna* wherein he clearly states that he obtained the formula from Ta'taglia and that the result was also found by del Ferro. For some reason the formula has become associated with Cardano's name, leading many to believe he stole it. But his book is widely available and the facts can be easily checked (see Note 1).

Now the equation $x^3 = 15x + 4$ was discussed by Bombelli in his algebra book written in 1572. Applying Ta'taglia's formula, he obtained the root

$$r = \sqrt[3]{2 + \sqrt{-121}} + \sqrt[3]{2 - \sqrt{-121}} \tag{1}$$

Bombelli wondered if, perhaps, the cube roots appearing in (1) might be written in the form $p + \sqrt{-q}$ and $p - \sqrt{-q}$, what we today would call complex conjugates (see Exercises 4, problem 1). He was able to show that they are indeed $2 + \sqrt{-1}$ and $2 - \sqrt{-1}$ (see Exercises 4, problem 10). Putting these in (1) he obtained $r = 4$, which is easily seen to satisfy the equation. Now expressions involving square roots of negative numbers had, on occasion, been considered before. In fact Cardano, in his book, used them to find two numbers whose sum is 10 and product is 40. But Bombelli went on to develop the arithmetic of these expressions, giving us the rules we use today. In modern notation he worked with quantities of the form $a + bi$ where a, b are real and $i^2 = -1$ and he developed the arithmetic operations on these quantities. Thus the system of complex numbers was born.

A major advance in our understanding of equations came in the 1620s when the English mathematician Thomas Harriot began writing them in the modern way, i.e., as a polynomial set equal to zero. So $x^3 = 15x + 4$ became $x^3 - 15x - 4 = 0$. Harriot noted that whenever the number r was a root of the equation $p(x) = 0$, he could find a polynomial $q(x)$ such that $p(x) = (x-r)q(x)$. Clearly, $q(x)$ must have degree one less than that of $p(x)$, and thus he discovered the fundamental fact that the degree of $p(x)$ tells us the number of roots of the equation $p(x) = 0$. By combining this with the work done in Italy, it became clear that every quadratic, every cubic, and every quartic (fourth-degree equations, first solved by Lodovico Ferrari, a student of Cardano's) had two, three, and four, respectively, complex solutions. The natural question then was whether one could show that every polynomial equation of degree n had n complex roots. Various people did make progress with this question, but complete success did not come until nearly 200 years later (1820) with the work of Gauss. Thus we see that complex numbers, despite their mysterious nature, became central to the theory of equations.

Still people were skeptical. Euler, who gave us the definition (generalizing an important observation of DeMoivre; see Exercises 4, problem 3b)

$$e^{iy} = \cos y + i \sin y, \quad y \in \mathbb{R} \tag{2}$$

wrote in 1770: "All such expressions as $\sqrt{-1}, \sqrt{-2}$, etc., are completely impossible or imaginary numbers, since they represent roots of negative quantities; and of such numbers we may truly assert that they are neither nothing, nor greater than nothing, nor less than nothing, which necessarily constitutes them imaginary or impossible."

The first clue as to how one might define complex numbers rigorously came from a geometric interpretation of these quantities given, independently, by a Norwegian surveyor named Wessel, who presented his ideas to the Danish Academy of Sciences in 1797; by the astronomer Bessel in 1799; and by a Swiss bookkeeper named Robert Argand in 1806. These results made the connection between complex numbers and points in the plane, i.e., ordered pairs of real numbers. Essentially what they did was to identify the symbol $a + bi$ with the point (a, b) or, if one preferred, with the segment joining $(0, 0)$ to (a, b). We shall exploit this representation below.

Let us turn now to Gauss' definition of the system of complex numbers first given in 1831. He defined his basic set, $\mathbb{C}$, to be the Cartesian product of $\mathbb{R}$ with itself, the set $\mathbb{R} \times \mathbb{R}$. Next he needed arithmetic operations on the members of $\mathbb{C}$ and these he defined as follows:

$$(a, b) + (c, d) = (a + c, b + d) \tag{3}$$
$$(a, b) \cdot (c, d) = (ac - bd, ad + bc) \tag{4}$$

for all $(a, b) \in \mathbb{C}$ and all $(c, d) \in \mathbb{C}$.

It is easy to see that $(0, 0)$ and $(1, 0)$ are, respectively, the additive and multiplicative identities of $\mathbb{C}$. Each complex number (a, b) has a unique additive inverse, $(-a, -b)$ of course, and if it is not zero, a unique multiplicative inverse (see Exercises 4, problem 1b). It is customary to identify $\mathbb{R}$ with the subset $\{(x, y) \in \mathbb{C} | y = 0\}$; thus every real number is also a complex number. Note that the real number -1 is now identified with $(-1, 0)$. Since $(0, 1) \cdot (0, 1) = (-1, 0)$, by (4), $(0, 1)$ is the mysterious i.

Now that we know what complex numbers really "are" we shall, as is customary, write them as $a + bi$, a and b real, $i^2 = -1$. Thus $\mathbb{C} = \{a + bi | a, b \in \mathbb{R}\}$. We may identify any $z = a + bi$ in $\mathbb{C}$ with the line segment joining $(0, 0)$ to (a, b). The length of this segment, $\sqrt{a^2 + b^2}$, is called the modulus, or absolute value, of z; we shall write $|z| = \sqrt{a^2 + b^2}$. It is clear from the geometry that we may write $a = |z| \cos \theta, b = |z| \sin \theta$ where θ is the angle between our segment and the positive x axis. The angle θ is referred to as an argument of z; obviously θ is not unique. Thus, for any $z = a + bi$ in $\mathbb{C}$, we have

$$z = a + bi = |z|(\cos\theta + i\sin\theta) = |z|e^{i\theta} \tag{5}$$

These last two expressions are called the polar forms of z. For any two points of $\mathbb{C}$, say $z_1 = x_1 + iy$ and $z_2 = x_2 + iy_2$, we define the distance between these points to be $|z_1 - z_2|$. This is just $\sqrt{(x_1 - x_2)^2 + (y_1 - y_2)^2}$ hence for many problems we may identify the complex plane with the plane of analytic geometry.

Exercises 4

1. For any complex number $z = a + bi$ define $\bar{z}$, read "z bar," to be $a - bi$. This is called the conjugate of z.

    *(a) Show that $z \cdot \bar{z} = a^2 + b^2 = |z|^2$.

    *(b) If $z \neq 0$, compute the multiplicative inverse of z, z^{-1}, and show that it is unique.

    (c) Define $f \colon \mathbb{C} \to \mathbb{C}$ by setting $f(z) = \bar{z}$ for all z. Show that (i) $f(z \cdot w) = f(z) \cdot f(w)$; (ii) $f(z + w) = f(z) + f(w)$; (iii) $f(az) = af(z)$ when a is real.

    (d) Use (c) to show that the complex roots of a polynomial equation having real coefficients must occur in conjugate pairs; i.e., if $p(z) = 0$, then we must also have $p(\bar{z}) = 0$.

2. Show that $e^{i\pi} + 1 = 0$. This formula was once thought to have mystical significance because it combines the e of analysis, the π of geometry, the zero and one of arithmetic, and the i of algebra.

*3. Let z and w be two complex numbers whose polar forms are, respectively, $z = r(\cos\theta + i\sin\theta)$, $w = s(\cos\phi + i\sin\phi)$.

    (a) Show that $z \cdot w = rs(\cos(\theta + \phi) + i\sin(\theta + \phi))$.

    (b) Prove DeMoivre's formula:
    For every natural number n, $z^n = r^n(\cos n\theta + i\sin n\theta)$.
    Hint: Use the fact that any nonempty set of natural numbers has a smallest member.

*4. Let z, w be any two complex numbers. Show that
    (a) $|z \cdot w| = |z| \cdot |w|$ and (b) $|z + w| \leq |z| + |w|$.

*5. Assume the fundamental theorem of algebra: Any polynomial equation, with complex coefficients, has a complex root. Use this to show that an equation of degree n has n complex roots.

*6. Given any function f from $\mathbb{R}$ to $\mathbb{C}$ show that there are two functions g and h, both from $\mathbb{R}$ to $\mathbb{R}$, such that $f(x) = g(x) + ih(x)$ for all $x \in \mathbb{R}$.

   Hint: Define $u : \mathbb{C} \to \mathbb{R}$ and $v : \mathbb{C} \to \mathbb{R}$ by setting, for each $z = ai + ib$, $u(z) = a$ and $v(z) = b$. Incidentally, a is called the *real part* of the complex number $a + bi$, and b is called its *imaginary part*. This custom is unfortunate because it forces us to live with the peculiar fact that the imaginary part of a complex number is real.

7. We refer here to section 3, specifically to the subset P of $\mathbb{R}$ with properties (5) and (6). Show that there is no subset of $\mathbb{C}$ that has these properties.

8. Describe geometrically the sets defined by the following inequalities: (a) $|z| < 1$; (b) $|z + i| < 1$; (c) $|z| \geq 2$; (d) $|z| = 1$.

9. Find the real part, the imaginary part (see Exercises 4, problem 6), and the absolute value of (a) $(3i - 2)^3$; (b) $\frac{i-1}{i+1}$; (c) $\frac{1}{2+3i}$.

10. Compute $(2 + i)^3$ and $(2 - i)^3$.

11. Show that for any natural number n the number i^n is equal to one, and of course only one, of the numbers i^0, i^1, i^2, i^3.

5. Analysis

The basic notions of analysis, limit, continuity, and differentiability are formally the same whether our functions map $\mathbb{R}$ to $\mathbb{R}$, $\mathbb{R}$ to $\mathbb{C}$, $\mathbb{C}$ to $\mathbb{R}$, or $\mathbb{C}$ to $\mathbb{C}$. To exploit this fact, we shall let $\mathbb{K}$ and $\mathbb{K}'$ denote either $\mathbb{R}$ or $\mathbb{C}$.

Definition 1. Let $s_0 \in \mathbb{K}$ and let ρ be a fixed, positive, real number. The set $U_\rho(s_0) = \{s \in \mathbb{K} \big| |s - s_0| < \rho\}$ is called an open neighborhood of s_0 or, sometimes, just a neighborhood of s_0. If $S \subseteq \mathbb{K}$ and $s_0 \in S$ we shall say that s_0 is an interior point of S if for some $\rho > 0$, $U_\rho(s_0) \subseteq S$. The set S is called an open set if each of its points is an interior point.

We leave it to the reader to show that ϕ and $\mathbb{K}$ are open sets and that (i) the union of any family of open sets is an open set, and (ii) the intersection of any finite family of open sets is an open set.

A subset of $\mathbb{K}$ is called a closed set if its complement is an open set. From (i) and (ii) and the theorem of DeMorgan (Exercises 2, problem 5) we immediately obtain (iii) the intersection of any family of closed sets is a closed set, and (iv) the union of any finite family of closed sets is a closed set.

A nonempty subset S of $\mathbb{K}$ is said to be bounded if there is a positive real number M such that $|s| \leq M$ for all $s \in S$. The empty set is also taken to be a bounded set. We recall that any set that is both closed and bounded is called a compact set. These behave very much like finite sets. In particular, when S is a compact set, we can, from any family of open sets whose union contains S, extract a finite subfamily whose union contains S.

If a function f is defined on $U_\rho(s_0)$, except possibly at s_0 itself, and if for some $t_0 \in \mathbb{K}$ we have: Given $\varepsilon > 0$ there is a $\delta > 0$ such that $|f(s) - t_0| < \varepsilon$ whenever $0 \neq |s - s_0| < \delta$, then we shall say that t_0 is the limit of $f(s)$ as s approaches s_0, and we shall write $\lim_{s \to s_0} f(s) = t_0$.

Definition 2. Let $f\colon S \to \mathbb{K}'$ and let s_0 be an interior point of S. We shall say that f is continuous at s_0 if $\lim_{s \to s_0} f(s) = f(s_0)$. We shall say that f is differentiable at s_0 if $\lim_{s \to s_0} \dfrac{f(s) - f(s_0)}{s - s_0}$ exists. This limit is denoted by $f'(s_0)$.

It is easy to see (see Exercises 5, problem 2) that if f is differentiable at s_0, it is continuous there. We stress that differentiability is defined only at interior points of the domain S of f. The following amusing example shows that this restriction is not to be taken lightly.

Example 1

One can easily show that for any natural number n,

$$(*) \quad n^2 = n + n + \cdots + n$$

where there are n summands on the right-hand side of $(*)$. Now the derivative of n^2 is $2n$ and that of n is 1, hence

$$(**) \quad 2n = 1 + 1 + \cdots + 1 = n$$

Dividing by n, which is valid because n is a natural number and hence not zero, we get $2 = 1$. ∎

Let us specialize now to partial functions from $\mathbb{R}$ to $\mathbb{R}$. More specifically, we consider a function f whose domain is a closed interval $[a, b]$. We already know what it means to say that f is continuous at any point $x_0 \in (a, b)$. We shall say that f is continuous at $x = a$ if given $\varepsilon > 0$ there is a $\delta > 0$ such that $|f(a) - f(x)| < \varepsilon$ whenever $x \in [a, b]$ and $x - a < \delta$. We shall say that f is continuous at $x = b$ if given $\varepsilon > 0$ there is a $\delta > 0$ such that $|f(b) - f(x)| < \varepsilon$ whenever $x \in [a, b]$ and $b - x < \delta$. Not surprisingly, we say that f is continuous on $[a, b]$ when it is continuous at each point of (a, b) and also continuous at $x = a$ and at $x = b$. Let us now review some of the basic properties of real-valued functions that are continuous on closed intervals (remember a, b are in $\mathbb{R}$ so the length of $[a, b]$ is finite). We shall omit the proofs.

Theorem 1. There are points x_1, x_2 of $[a, b]$ such that $f(x_1) \leq f(x) \leq f(x_2)$ for all x in $[a, b]$.

This theorem simply states that f has a maximum and a minimum on $[a, b]$. Suppose now that y is any number that is between the two extremes of f. Then:

Theorem 2. There is a point $x_0 \in [a, b]$ such that $f(x_0) = y$.

Definition 3. Let $f\colon [a, b] \to \mathbb{K}'$. We shall say that f is uniformly continuous on $[a, b]$ if, given $\varepsilon > 0$, there is a $\delta > 0$ such that $|f(x_1) - f(x_2)| < \varepsilon$ whenever x_1, x_2 are in $[a, b]$ and $|x_1 - x_2| < \delta$.

Theorem 3. If $f\colon [a, b] \to \mathbb{R}$ is continuous on $[a, b]$, then it is uniformly continuous on this interval.

Theorem 4. If $f\colon [a, b] \to \mathbb{R}$ is continuous on $[a, b]$, then the Riemann integral of f, $\int_a^b f(x)dx$ exists. Moreover, there is a point $x_0 \in [a, b]$ such that $\int_a^b f(x)dx = f(x_0)(b - a)$.

The second statement is an immediate consequence of theorems 1 and 2. Finally, we have the mean-value theorem:

Theorem 5. If $f\colon [a, b] \to \mathbb{R}$ is continuous and has a derivative at each point of (a, b), then to any h such that $a < a + h < b$ there corresponds a number $\theta, 0 < \theta < 1$ such that $f(a + h) - f(a) = hf'(a + \theta h)$.

Let us turn now to functions from $\mathbb{C}$ to $\mathbb{C}$. Simple examples are as follows:

$$f(z) = z^2 = (x + iy)^2 = (x^2 - y^2) + i(2xy) \tag{1}$$

$$f(z) = e^z = e^{x+iy} = e^x(\cos y + i \sin y) = e^x \cos y + i(e^x \sin y) \tag{2}$$

It is clear, as we have illustrated, that any $f\colon \mathbb{C} \to \mathbb{C}$ can be written in the form

$$f(z) = u(x, y) + iv(x, y) \tag{3}$$

where u and v are real-valued functions defined on $\mathbb{R}^2$. Suppose now that f is differentiable at $z_0 = x_0 + iy_0$. Then f is defined on some neighborhood $U_\rho(z_0)$ and

$$\lim_{z \to z_0} \frac{f(z) - f(z_0)}{z - z_0} = f'(z_0)$$

We must obtain the same number for this limit no matter how z approaches z_0. Let us take, first, $z = (x_0 + \Delta x) + iy_0$, where Δx is so small that $z \in U_\rho(z_0)$. Then we find that

$$f'(z_0) = \lim_{\Delta x \to 0} \frac{u(x_0 + \Delta x, y_0) - u(x_0, y_0)}{\Delta x} + i \lim_{\Delta x \to 0} \frac{v(x_0 + \Delta x, y_0) - v(x_0, y_0)}{\Delta x}$$
$$= \frac{\partial u}{\partial x}(x_0, y_0) + i\frac{\partial v}{\partial x}(x_0, y_0)$$

But by taking $z = x_0 + i(y_0 + \Delta y)$ we obtain, similarly,

$$f'(z_0) = -i\frac{\partial u}{\partial y}(x_0, y_0) + \frac{\partial v}{\partial y}(x_0, y_0) \tag{4}$$

Thus at any point at which f has a derivative, we have

$$\frac{\partial u}{\partial x} = \frac{\partial v}{\partial y} \tag{5}$$

and

$$\frac{\partial u}{\partial y} = -\frac{\partial v}{\partial x} \tag{6}$$

These are the Cauchy-Riemann equations named for the Frenchman A. L. Cauchy and the German G. F. B. Riemann, two pioneers in complex function theory.

It is clear that $u(x,y) = x^2 - y^2$, $v(x,y) = 2xy$ satisfy these equations as do $u(x,y) = e^x \cos y$ and $v(x,y) = e^x \sin y$.

Suppose now that we have two real-valued functions u and v defined on $\mathbb{R}^2$ that satisfy the Cauchy-Riemann equations at some point. Suppose further that u and v have continuous first and second partial derivatives at this point. Then equations (5) and (6) give:

$$\frac{\partial^2 u}{\partial x^2} = \frac{\partial^2 v}{\partial x \partial y} \tag{7}$$

and

$$\frac{\partial^2 u}{\partial y^2} = -\frac{\partial^2 v}{\partial y \partial x} \tag{8}$$

Under our assumptions, $\frac{\partial^2 v}{\partial x \partial y} = \frac{\partial^2 v}{\partial y \partial x}$, hence u must satisfy

$$\frac{\partial^2 u}{\partial x^2} + \frac{\partial^2 u}{\partial y^2} = 0 \tag{9}$$

A similar calculation shows that v must satisfy the same equation. Equation (9) is called Laplace's equation or, sometimes, the potential equation. We have already seen four solutions to this equation, and our discussion above shows how we can generate many more (see Exercises 5, problem 4). This is a very important observation as we shall see.

A function $g\colon \mathbb{N} \to S$ (recall that $\mathbb{N} = \{1, 2, 3, \dots\}$), where S is an arbitrary set, is called a sequence of points of S, or, more simply, a sequence in S. It is customary to set $g(n) = s_n$ for all n and to speak of the sequence $\{s_n\}_{n=1}^{\infty}$.

Definition 4. Let $\{s_n\}_{n=1}^{\infty}$ be a sequence in $\mathbb{K}$. We shall say that $\{s_n\}$ is convergent to the element $s_0 \in \mathbb{K}$, and we shall write $\lim_{n \to \infty} s_n = s_0$, if given $\varepsilon > 0$ there is a natural number n_0 such that $|s_0 - s_n| < \varepsilon$ whenever $n \geq n_0$. A sequence that is not convergent is said to be divergent.

When f is a partial function from $\mathbb{K}$ to $\mathbb{K}'$ and $\{s_n\}$ is a sequence in the domain of f, then $\{f(s_n)\}$ is a sequence in the range of f. We can use the behavior of sequences constructed in this way to characterize those points at which f is continuous.

Theorem 6. Let f be a partial function from $\mathbb{K}$ to $\mathbb{K}'$ and let s_0 be an interior point of the domain of f. Then f is continuous at s_0 if, and only if, $\lim_{n \to \infty} f(s_n) = f(s_0)$ whenever $\{s_n\}$ is a sequence in the domain of f that converges to s_0.

Proof. Suppose, first, that f is continuous at s_0 and that $\{s_n\}$ is a sequence in the domain of f that converges to s_0. Given $\varepsilon > 0$, we use the continuity of f to find $\delta > 0$ such that $|f(s_0) - f(s)| < \varepsilon$ whenever $|s - s_0| < \delta$; remember s_0 is an interior point of the domain of f. Next we use the fact that $\lim_{n \to \infty} s_n = s_0$ to find a natural number n_0 such that $|s_n - s_0| < \delta$ whenever $n \geq n_0$. Thus for the given $\varepsilon > 0$, we have $|f(s_n) - f(s_0)| < \varepsilon$ whenever $n \geq n_0$, and this means $\{f(s_n)\}$ is convergent to $f(s_0)$ as claimed.

Now suppose that $\lim f(s_0) = f(s_0)$ whenever $\{s_n\}$ converges to s_0 and yet f is not continuous at this point. We shall deduce a contradiction. Negating the definition of continuity at s_0, we see that for some $\varepsilon > 0$ we can, for every $\delta > 0$, find an element s_δ in the domain of f for which $|s_\delta - s_0| < \delta$ and yet $|f(s_\delta) - f(s_0)| \geq \varepsilon$. We now construct a sequence as follows. Setting $\delta = 1$ we find s_1 such that $|s_1 - s_0| < 1$ and $|f(s_1) - f(s_0)| \geq \varepsilon$. Next we set $\delta = \frac{1}{2}$ and find s_2 so that $|s_2 - s_0| < \frac{1}{2}$ and $|f(s_2) - f(s_0)| \geq \varepsilon$. We continue in this way. After $s_1, s_2, \ldots, s_{n-1}$ have been found, we set $\delta = \frac{1}{n}$ and find s_n so that $|s_n - s_0| < \frac{1}{n}$ and $|f(s_n) - f(s_0)| \geq \varepsilon$. It is clear that we have constructed a sequence $\{s_n\}$ that converges to s_0. But it is also clear that $\{f(s_n)\}$ could not converge to $f(s_0)$ because $|f(s_n) - f(s_0)| \geq \varepsilon$ for all natural numbers n. This is our contradiction.

Definition 5. A sequence $\{s_n\} \subseteq \mathbb{K}$ is called a Cauchy sequence if, given any $\epsilon > 0$, there is a natural number n_0 such that $|s_n - s_m| < \epsilon$ whenever m and n both exceed n_0.

If $\{s_n\}$ converges to s_0, then because $|s_n - s_m| \leq |s_n - s_0| + |s_0 - s_m|$ we see that this sequence must be a Cauchy sequence. For the set $\mathbb{R}$, it is in fact the case that a sequence converges if, and only if, it is a Cauchy sequence. This statement is equivalent to the completeness axiom for $\mathbb{R}$ (Section 3, paragraph just after equation (8)). It follows easily that a sequence in $\mathbb{C}$ is convergent if, and only if, it is a Cauchy sequence (see Exercises 5, problem 1).

Exercises 5

1. Let $\{z_n\}$ be a sequence in $\mathbb{C}$, and for each n let $z_n = x_n + iy_n$, where x_n and y_n are in $\mathbb{R}$.

    (a) Show that $\{z_n\}$ is a Cauchy sequence if, and only if, both $\{x_n\}$ and $\{y_n\}$ are Cauchy sequences.

    (b) Show that $\lim_{n \to \infty} z_n = x_0 + iy_0$ if, and only if, $\lim x_n = x_0$ and $\lim y_n = y_0$.

*2. If the function f is differentiable at a point $s_0 \in \mathbb{K}$, show that it is continuous at this point.

*3. Suppose $f \colon \mathbb{R} \to \mathbb{C}$. Then we may write $f(x) = g(x) + ih(x)$, where g and h map $\mathbb{R}$ to $\mathbb{R}$.

    (a) Show that $\lim_{x \to x_0} f(x) = u_0 + iv_0$ if, and only if, we have both $\lim_{x \to x_0} g(x) = u_0$ and $\lim_{x \to x_0} h(x) = v_0$.

    (b) Show that f is continuous on $[a,b]$ if, and only if, both g and h are continuous on this interval. So $\int_a^b f(x)dx$ can be defined to be $\int_a^b g(x)dx + i\int_a^b h(x)dx$, and we shall do so.

    (c) Show that f is differentiable at a point x_0 if, and only if, both g and h are differentiable at this point. Show that, when this is the case, $f'(x_0) = g'(x_0) + ih'(x_0)$.

4. (a) The function $f(z) = z^3$ is differentiable at every point in $\mathbb{C}$. Find the real and imaginary parts of this function and show that together they satisfy the Cauchy-Riemann equations. Show that each of these functions satisfies Laplace's equations.

    (b) Do the real and imaginary parts of the function $f(z) = \bar{z}$ satisfy the Cauchy-Riemann equations?

5. Let f be a partial function from $\mathbb{R} \to \mathbb{K}'$ with domain a closed interval $[a,b]$. Show that f is continuous on $[a,b]$ if, and only if, we have (i) f is continuous on (a,b); (ii) $\lim f(x_n) = f(a)$ whenever $\{x_n\} \subseteq [a,b]$ and $\lim x_n = a$; (iii) $\lim f(x_n) = f(b)$ whenever $\{x_n\} \subseteq [a,b]$ and $\lim x_n = b$.

6. Let f be a partial function from $\mathbb{K}$ to $\mathbb{K}'$ that is uniformly continuous (Definition 3) on its domain.

    (a) Show that $\{f(s_n)\}$ is a Cauchy sequence whenever $\{s_n\} \subseteq \mathcal{D}(f)$ is a Cauchy sequence.

    (b) Show that $f(x) = \frac{1}{x}$, $\mathcal{D}(f) = (0,1)$, is not uniformly continuous.

*7. Let $\{n_k\}_{k=1}^{\infty}$ be a sequence of natural numbers such that $n_k < n_{k+1}$ for all k. Given a sequence $\{s_n\}$, the sequence $\{s_{n_k}\}$ is called a subsequence of $\{s_n\}$.

(a) Show that all subsequences of a convergent sequence are convergent to the same limit.

(b) Show that any subsequence of a Cauchy sequence is a Cauchy sequence.

(c) If $\{s_n\}$ is a Cauchy sequence and if it has a subsequence that converges, show that $\{s_n\}$ is convergent.

(d) If $\{s_n\}$ is a Cauchy sequence, show that there is a positive real number M such that $|s_n| \leq M$ all n.

Notes

1. Cardano's book, *Ars Magna*, published in 1545, has been translated by T. Richard Witmer (*The Great Art*. M.I.T. Press, Cambridge, MA, 1968). On page 8 Cardano states that the formula for solving a cubic was found by Scipione del Ferro and, independently, by Ta'taglia. He also states that the formula was given to him by Ta'taglia. Nowhere does he claim it as his own.

2. Bombelli's work, and a comment from him on how he was led to these ideas, can be found in *Number—the Language of Science*, by Tobias Dantzig (Doubleday and Co., Inc., Garden City, NY, 4th edition, 1956). He discusses Bombelli's derivation of the laws of complex arithmetic and also the work of Harriot (see pp. 184–186 and p. 188). Euler's quote about the impossibility of taking square roots of negative numbers can be found on page 191. The geometric interpretation of complex numbers by Wessel and later by Argand is discussed on page 101.

3. Ralph P. Boas, in his wonderful book, *Invitation to Complex Analysis* (Random House, Inc., New York, 1987), states that the geometric interpretation of complex numbers was also found by Bessel in 1799 (see the notes on p. 13).

Chapter One
Classical Harmonic Analysis

The central question in this subject first arose in the 18th century when D'Alembert, Euler, and Bernoulli investigated the problem of the vibrating string. We imagine an elastic string, of finite length, stretched along the x-axis with both end points fixed. If the string is displaced to form a curve, say $f(x)$, and then released, the y-coordinate of each of its points will vibrate (i.e., vary with time), and because the end points are fixed, y depends on x. So we'll have

$$y = y(x, t) \tag{1}$$

with $y(x,0) = f(x)$ and $f(x)$ is known. It was soon discovered that the function $y(x,t)$ could be found provided one could represent $f(x)$ as a series of sines and cosines, i.e., as a trigonometric series. More explicitly, provided one could find constants $a_0, a_1, a_2, \cdots$ and $b_1, b_2, \cdots$ such that

$$f(x) = \frac{1}{2}a_0 + \sum_{n=1}^{\infty}(a_n \cos nx + b_n \sin nx) \tag{2}$$

The details of the analysis connecting the physical problem and the series representation are discussed later.

Many writers begin their discussions of classical harmonic analysis with the series (2). We believe, however, that it is more instructive to stay closer to the historical development. So we shall begin with a problem of mathematical physics. A problem whose solution is, we believe, especially illuminating.

1. The Dirichlet Problem for a Disk

Before we can state the problem we have in mind we need some terminology.

Definition 1. Let C be a circle in the plane (i.e., $\mathbb{R}^2$) and let G be the set of all points inside C. A real-valued function u defined in G is said to be harmonic in G if (i) u has continuous partial derivatives of order one and of order two at every point of this set, and (ii) at each point of our set u satisfies the equation $\frac{\partial^2 u}{\partial x^2} + \frac{\partial^2 u}{\partial y^2} = 0$.

The equation just written is called the potential equation or, sometimes, Laplace's equation. In Chapter Zero (Section 5) we gave examples of harmonic functions and indicated how one can find many others. These functions have some remarkable properties.

Theorem 1. Let C be a circle of diameter d in $\mathbb{R}^2$ and let G be the set of all points inside C. Suppose that u is a real-valued function that is defined and continuous on $G \cup C$ and is harmonic in G. Then the values of u in G cannot exceed the maximum of u on C, nor can they be less than the minimum of u on C.

Proof. Let m be the maximum of u on C, and suppose that at some point $(x_0, y_0) \in G$, $u(x_0, y_0) = M > m$. We may suppose that M is the maximum of u on $G \cup C$. Now translate the origin to (x_0, y_0). We obtain a function we shall denote by u that has all of the properties of our original function and has maximum m over C and maximum M over $G \cup C$ but attains this last value at $(0,0)$. Now define an auxiliary function $v(x,y) = u(x,y) + \frac{M-m}{2d^2}(x^2 + y^2)$. Clearly, $v(0,0) = M$ and, if $(x,y) \in C$, then $v(x,y) \leq m + \left(\frac{M-m}{2d^2}\right)d^2 = \frac{M+m}{2} < M$.

Thus v attains its maximum over $G \cup C$ at some point of G. But, at every point of G

$$\frac{\partial^2 v}{\partial x^2} + \frac{\partial^2 v}{\partial y^2} = \frac{\partial^2 u}{\partial x^2} + \frac{\partial^2 u}{\partial y^2} + 2\left(\frac{M-m}{d^2}\right) = 2\frac{M-m}{d^2} > 0$$

This is a contradiction because, at the maximum of v, none of its second partials can be positive.

To prove the second statement of the theorem we apply the result just obtained to the function $-u(x,y)$. This proof is due to the Russian mathematician I. I. Privalov.

The problem we shall investigate here, and in the remainder of this chapter, is this: Let C be a circle in $\mathbb{R}^2$ and let G be the set of all points inside C. Suppose that we are given a real-valued function f that is defined and continuous on C. Can we find a function u that is harmonic in G, continuous on $G \cup C$, and equal to f on C?

This is called the Dirichlet problem (or, simply, the DP) for C with boundary data f. Theorem 1 has an immediate and, as we shall see, very important, consequence.

Theorem 2. Solutions to the Dirichlet problem for a circle are unique.

Proof. Using the notation introduced above, suppose that u_1 and u_2 are two solutions to the DP for the disk $G \cup C$ with boundary data f. Then the function $u_1 - u_2$ is harmonic in G, continuous on $G \cup C$, and identically zero on C. By Theorem 1 this function must be identically zero at all points of G. Thus $u_1 = u_2$ on $G \cup C$.

Exercises 1

*1. We shall solve the DP for the unit disk, i.e., the disk whose boundary is the circle of radius one centered at the origin. To do that it is convenient to use polar coordinates: $x = r\cos\theta$, $y = r\sin\theta$. Show that under this transformation, $\dfrac{\partial^2 u}{\partial x^2} + \dfrac{\partial^2 u}{\partial y^2} = 0$ becomes $\dfrac{\partial^2 u}{\partial r^2} + \dfrac{1}{r}\dfrac{\partial u}{\partial r} + \dfrac{1}{r^2}\dfrac{\partial^2 u}{\partial \theta^2} = 0$.

2. Let C be a circle in the plane and let G be the set of all points inside C. Suppose that f_1, f_2 are two real-valued functions that are defined and continuous at each point of C and that these functions differ by less than some $\varepsilon > 0$ at each such point. If u_1 and u_2 are solutions to the (DP) for $G \cup C$ with boundary data f_1 and f_2, respectively, show that u_1 and u_2 differ by less than ε at each point of G.

2. Continuous Functions on the Unit Circle

The boundary data for the DP are continuous functions on a circle. Such functions have some interesting properties that we shall explore here. Recall that $\mathbb{K}$ denotes either the set of real numbers, $\mathbb{R}$, or the set of complex numbers, $\mathbb{C}$.

Definition 1. For any function $h\colon \mathbb{R} \to \mathbb{K}$, the set $P(h) = \{\alpha \in \mathbb{R} \mid h(x+\alpha) = h(x) \text{ for all } x\}$ is called the set of periods of h. We shall say that h is a periodic function when this set contains a nonzero number.

For any function h the set $P(h)$ contains the number zero, and in certain cases it contains nothing else. For example, when $h(x) = x$ or $h(x) = e^x$, we have $P(h) = \{0\}$. At the other extreme, we have $P(h) = \mathbb{R}$ when $h(x)$ is a constant; so constant functions are periodic. Less trivial examples are the functions $\sin x$ and $\cos x$. We recall that

$$P(\sin x) = P(\cos x) = \{2n\pi \mid n \in \mathbb{Z}\}$$

The set of periods of a function has some structure.

Lemma 1. For any function $h\colon \mathbb{R} \to \mathbb{K}$ we have (i) the number $(-\alpha) \in P(h)$ whenever $\alpha \in P(h)$, and (ii) if α, β are in $P(h)$, so are both the numbers $\alpha + \beta, \alpha - \beta$.

Proof. When $\alpha \in P(h)$ we have $h(x+\alpha) = h(x)$ for all x. Hence, when $x = y-\alpha$ we have $h(y) = h(y-\alpha)$, and this is true for any y.

Given α, β in $P(h)$ we see that $h[x+(\alpha+\beta)] = h[(x+\alpha)+\beta] = h(x+\alpha) = h(x)$ for all x, first because $\beta \in P(h)$ and then because α does. The fact that $(\alpha - \beta)$ is in $P(h)$ follows from what we have just shown.

Corollary 1. If $\alpha \in P(h)$, then $\{n\alpha \mid n \in \mathbb{Z}\} \subseteq P(h)$. When α is the smallest positive member of $P(h)$, these two sets coincide.

Proof. To show that $P(h) \supseteq \{n\alpha \mid \alpha \in \mathbb{Z}\}$, it is sufficient to show that $P(h) \supseteq \{n\alpha \mid n \in \mathbb{N}\}$. If this last statement is false, then $\{n \in \mathbb{N} \mid n\alpha \notin P(h)\}$ is nonempty, and hence it contains a smallest member n_0. From lemma 1 we see that $n_0 \geq 2$, and so $(n_0 - 1) \in \mathbb{N}$. Furthermore, $(n_0 - 1)\alpha$ must be in $P(h)$. But then, again using lemma 1, $(n_0 - 1)\alpha + \alpha = n_0\alpha$ is in $P(h)$, which is a contradiction.

Now suppose that $P(h)$ has a smallest positive element α. We want to show that $P(h) = \{n\alpha \mid n \in \mathbb{Z}\}$. Given any $\beta \in P(h)$ suppose, first, that $\beta > 0$. Then $\alpha \leq \beta$ and we can find an integer k such that $\beta = k\alpha + r$, $0 \leq r < \alpha$. But both β and $k\alpha$ are in $P(h)$, hence $r \in P(h)$, and so this number must be zero. Thus $\beta = k\alpha$, $k \in \mathbb{N}$. When β is negative, then $(-\beta) \in P(h)$, and this number is positive. So, $(-\beta) = k\alpha$ for $k \in \mathbb{N}$, telling us that $\beta = (-k)\alpha$.

A constant function is, obviously, a periodic function with no smallest positive period. A less trivial example is given in Exercises 2 (problem 2).

When a periodic function h has a smallest positive period, this number is sometimes called the period of h, or the Period of h. We shall call every element of $P(h)$ a period of h, and when we need to talk about the smallest positive member of $P(h)$ we shall refer to it as such.

Lemma 2. If a continuous function on $\mathbb{R}$ has arbitrarily small periods, it is a constant.

Proof. We shall prove this result for real-valued functions and leave the case of complex valued functions for the exercises (see Exercises 2, problem 6d).

So let $f : \mathbb{R} \to \mathbb{R}$ be a continuous function and let $\{\alpha_n\} \subseteq P(f)$ be positive numbers such that $\lim_{n \to \infty} \alpha_n = 0$. Now for each fixed n and any real number x, there is, by the mean-value theorem for integrals (Chapter Zero), a real number x_n^* such that

$$\int_x^{x+\alpha_n} f(t)dt = f(x_n^*)[(x+\alpha_n) - x] = f(x_n^*)\alpha_n$$

and $x \leq x_n^* \leq x + \alpha_n$. Thus

$$\frac{1}{\alpha_n} \int_x^{x+\alpha_n} f(t)dt = f(x_n^*)$$

Now f is continuous and $\alpha_n \to 0$, hence

$$\lim_{n \to \infty} \frac{1}{\alpha_n} \int_x^{x+\alpha_n} f(t)dt = f(x)$$

However, because $\alpha_n \in P(f)$ we have, for any x and y,

$$\int_x^{x+\alpha_n} f(t)dt = \int_y^{y+\alpha_n} f(t)dt$$

(see Exercises 2, problem 3). Thus letting $n \to \infty$, we find that $f(x) = f(y)$ showing that f is a constant.

Corollary 2. Any nonconstant, continuous, periodic function has a smallest positive period.

Proof. Again we shall prove our result only for real-valued functions. So let $f \colon \mathbb{R} \to \mathbb{R}$ be nonconstant, continuous, and periodic. By lemma 1, the set $P_+(f) = \{\alpha \in P(f) \mid \alpha > 0\}$ is nonempty. Let α_0 be the greatest lower bound for this set (Chapter Zero). If α_0 is zero, then a sequence in $P_+(f)$, a sequence of positive periods of f, must converge to it; i.e., f would have arbitrarily small periods. This cannot be the case, because f is not constant, and so α_0 must be positive.

To complete the proof we need only show that $\alpha_0 \in P_+(f)$. We argue by contradiction. If α_0 is not in this set, then we must have $\{\alpha_n\} \subseteq P_+(f)$ such that $\lim_{n \to \infty} \alpha_n = \alpha_0$. But then, because f is continuous, we have $f(x+\alpha_0) = \lim_{n \to \infty} f(x+\alpha_n)$ for any fixed x. However, $f(x+\alpha_n) = f(x)$, so $f(x+\alpha_0) = f(x)$ for any x, and this says $\alpha_0 \in P(f)$.

Combining our results we see that if f is a nonconstant, continuous, periodic function, then f has a smallest positive period α_0 and $P(f) = \{n\alpha_0 \mid n \in \mathbb{Z}\}$.

In all that follows we shall denote the unit circle (the circle of radius one centered at the origin) by T. We shall solve the DP for this circle. Note that if f is a continuous function on T, then we may define $\tilde{f}(\theta)$ to be $f(\cos\theta, \sin\theta)$ for all $\theta \in \mathbb{R}$. It is clear that $\widetilde{f}$ is a continuous function on $\mathbb{R}$ and that 2π is a period of this function. Thus every continuous function on T gives us a continuous function on $\mathbb{R}$ that has 2π for a period. Conversely, given a continuous function $\widetilde{f}$ on $\mathbb{R}$ that has 2π for a period, we may define $f(\cos \theta, \sin \theta)$ to be $\widetilde{f}(\theta)$, for all θ, obtaining a function f that is continuous on T.

Definition 2. A function that has 2π for a period, whether or not it is the smallest period of this function, is called a 2π-periodic function.

Exercises 2

1. Let $f \colon \mathbb{R} \to \mathbb{K}$. Show that $P(f) = \mathbb{R}$ if, and only if, f is constant.

2. Recall the set of rational numbers Q. Define $\chi(x) = 1$ when $x \in Q$, $\chi(x) = 0$ when $x \in \mathbb{R} \setminus Q$. Show that $P(\chi) = Q$; hence, χ has no smallest positive period.

* 3. Let $f \colon \mathbb{R} \to \mathbb{R}$ be continuous and periodic and let $\alpha > 0$ be any element of $P(f)$. Show that for any real numbers x and y we have $\int_x^{x+\alpha} f(t)dt = \int_y^{y+\alpha} f(t)dt$.

* 4. Let a, b be two fixed complex numbers and let λ be a positive real number. Let $g(x) = a\cos \lambda x + b\sin \lambda x$. Show that $P(g) = \{\frac{2n\pi}{\lambda} | n \in Z\}$.

5. Let f, g be two periodic functions.

    (a) If the set $P(f) \cap P(g)$ contains a nonzero element, show that $af + bg$ (here a, b are any complex constants) and $f \cdot g$ are periodic functions.

    (b) When f has a smallest positive period α and g has a smallest positive period β, show that $P(f) \cap P(g)$ contains a nonzero number if, and only if, $\alpha = r\beta$ for some rational number $r \neq 0$.

* 6. Let $h \colon \mathbb{R} \to \mathbb{C}$

    (a) Show that there are functions f, g, both from $\mathbb{R}$ to $\mathbb{R}$, such that $h(x) = f(x) + ig(x)$ for all x.

    (b) Show that $P(h) = P(f) \cap P(g)$. Hence if h is periodic, then both f and g are periodic; the converse is false.

    (c) Show that h is continuous if, and only if, both f and g are continuous.

    (d) If h is continuous and has arbitrarily small periods, show that h is a constant.

    (e) If h is continuous, periodic, and nonconstant, show that h has a smallest positive period.

* 7. If $f \colon \mathbb{R} \to \mathbb{C}$ is continuous and periodic, show that it is uniformly continuous on $\mathbb{R}$ (Chapter Zero, Section 5, Definition 3).

3. The Method of Fourier

The DP requires that we solve a certain partial differential equation subject to a given boundary condition. The mathematical details are a bit complicated, and it is wise to begin with a simpler problem involving the solution of an ordinary differential equation. The relevance of this simpler problem will become clear very soon.

We begin with the equation

$$\Theta''(\theta) + \mu\Theta(\theta) = 0, \qquad \mu \text{ a real constant} \tag{1}$$

We seek nontrivial (i.e., not identically zero) solutions to (1) that are 2π-periodic. There are three cases:

(i) $\mu = 0$. Here the general solution to our equation ($\Theta''(\theta) = 0$) is easily seen to be

$$\Theta(\theta) = c_0\theta + a_0$$

where c_0 and a_0 are constants. This is nontrivial and 2π-periodic only when c_0 is zero and a_0 is not.

(ii) $\mu < 0$, say $\mu = -\lambda^2$ where λ is a positive real number. So (1) becomes

$$\Theta''(\theta) - \lambda^2\Theta(\theta) = 0$$

We set $\Theta(\theta) = e^{m\theta}$, m to be determined, and putting this into our equation, we find that $(m^2 - \lambda^2)e^{m\theta} = 0$. It follows that $m = \pm\lambda$ and the general solution to our equation is

$$\Theta(\theta) = c_1 e^{\lambda\theta} + c_2 e^{-\lambda\theta}$$

Clearly, there is no choice of the constants c_1 and c_2 that gives us a function that is both nontrivial and 2π-periodic. The problem has no solutions when $\mu < 0$ (see Exercises 3, problem 1).

(iii) $\mu > 0$, let us say $\mu = \lambda^2$ for some $\lambda > 0$. Again putting $\Theta(\theta) = e^{m\theta}$ into our equation (i.e., into $\Theta''(\theta) + \lambda^2\Theta(\theta) = 0$) we find that $m = \pm i\lambda$. Hence, the general solution is

$$\Theta_\lambda(\theta) = c_{1\lambda} e^{i\lambda\theta} + c_{2\lambda} e^{-i\lambda\theta} = a_\lambda \cos\lambda\theta + b_\lambda \sin\lambda\theta$$

where a_λ, b_λ are arbitrary constants, not both zero. We are seeking solutions that are 2π-periodic. Now

$$P(\Theta_\lambda) = \{n\left(\frac{2\pi}{\lambda}\right) \mid n \in \mathbb{Z}\} \qquad \text{(Exercises 2, problem 4)}$$

and so this condition is satisfied if, and only if, λ is a natural number.

To sum up, equation (1) has nontrivial 2π-periodic solutions when, and only when, $\mu = n^2$, $n = 0, 1, 2, \cdots$ and these are

$$\Theta_n(\theta) = a_n \cos n\theta + b_n \sin n\theta \qquad (2)$$

Let us turn now to the DP for the unit circle T. The geometry suggests that we use polar coordinates, and we let G be the set of all points inside T. Thus we are given a real-valued function $f(\theta)$ that is continuous on T, and we seek a function $u(r, \theta)$ that satisfies

$$\frac{\partial^2 u}{\partial r^2} + \frac{1}{r}\frac{\partial u}{\partial r} + \frac{1}{r^2}\frac{\partial^2 u}{\partial \theta^2} = 0 \qquad \text{(Exercises 1, problem 1)} \qquad (3)$$

at all points of G, that is continuous on $G \cup T$, and is equal to $f(\theta)$ on T.

Now equation (1) has, depending on μ, a general solution that can be represented by a simple formula. Equation (3), however, has so many particular solutions (Chapter Zero) that we cannot expect to represent all of them in any simple way. Another approach is needed, and the standard one is to restrict ourselves to solutions $u(r, \theta)$ that can be written in the form $R(r)\Theta(\theta)$. This may seem like a drastic assumption and a severe limitation on what we might find. One might wonder if there are any solutions that can be "factored" in this way. And, even if there are such solutions, might there not be other perhaps more interesting solutions that cannot be so written? Are we to ignore these? Both objections disappear, however, once we show that our restriction does in fact lead us to a solution, because, as we have seen (theorem 2 of Section 1), solutions to the DP for a circle are unique. This is what justifies our entire approach.

So let us set $u(r, \theta) = R(r)\Theta(\theta)$ and put this into (3). We find, after some algebra,

$$\frac{r^2 R'' + r R'}{R} = -\frac{\Theta''}{\Theta}$$

Because r and θ are independent variables, it follows that both expressions are equal to the same constant, say, μ. Thus we have two ordinary differential equations

$$r^2 R''(r) + r R'(r) - \mu R(r) = 0 \qquad (4)$$

and

$$\Theta''(\theta) + \mu \Theta(\theta) = 0 \qquad (5)$$

For each fixed μ we find the general solution to (4) and to (5) and their product is a solution to (3). However, we want $u(r, \theta) = R(r)\Theta(\theta)$ to be equal to the given function $f(\theta)$ when $r = 1$, i.e., on T. This says that the solutions to (5) that are of interest here must be 2π-periodic. Our discussion at the beginning of this section shows that we will find such solutions when, and only when, $\mu = n^2$ where $n = 0, 1, 2, \cdots$ and that these solutions are

$$\Theta_n(\theta) = a_n \cos n\theta + b_n \sin n\theta \tag{6}$$

Equation (4) now becomes

$$r^2 R''(r) + r R'(r) - n^2 R(r) = 0 \tag{7}$$

and its general solution, when $n \neq 0$, is easily seen to be

$$R(r) = c_1 r^n + c_2 r^{-n} \tag{8}$$

(see Exercises 3, problem 2). Because $R(r)\Theta(\theta)$ is to be harmonic inside the unit circle, and $c_2 r^{-n}$ is undefined at the origin, we must take $c_2 = 0$. When $n = 0$, the general solution to (7) is also undefined at the origin and hence cannot be harmonic (see Exercises 3, problem 3).

We have found a whole class of solutions to (3) that are 2π-periodic when $r = 1$. They are

$$r^n(a_n \cos n\theta + b_n \sin n\theta), \qquad n = 0, 1, 2, 3, \cdots \tag{9}$$

It would be remarkable if any one of these solutions was equal to the given function $f(\theta)$ on T. However, because (3) is linear (meaning sums and scalar multiples of solutions to this equation are again solutions), the early mathematicians simply "added" them all up to obtain

$$u(r, \theta) = a_0 + \sum_{n=1}^{\infty} r^n(a_n \cos n\theta + b_n \sin n\theta) \tag{10}$$

and then asked if one could choose the constants a_n and b_n so that

$$f(\theta) = u(1, \theta) = a_0 + \sum_{n=1}^{\infty} (a_n \cos n\theta + b_n \sin n\theta) \tag{11}$$

This is the problem that so troubled D'Alembert, Euler, and Bernoulli. Modern mathematicians have learned to be wary of infinite sums. We immediately ask about the meaning of (11). Some form of convergence is implied, but convergence in what sense?

Exercises 3

1. Let λ be a fixed, positive, real number.

   (a) Show that $ae^{\lambda x} + be^{-\lambda x} = 0$ for all x if, and only if, $a = b = 0$.

(b) Set $f(x) = c_1 e^{\lambda x} + c_2 e^{-\lambda x}$ where c_1, c_2 are constants not both zero. Show that $P(f) = \{0\}$, i.e., if $f(x+\alpha) = f(x)$ for all x, then $\alpha = 0$.

*2. Use the substitution $r = e^t$ to show that equation (7) is equivalent to $\ddot{R} - n^2 R = 0$ where $\dot{R} = \frac{d}{dt} R$. Next find the general solution to (7).

*3. Solve the equation $r^2 R''(r) + r R'(r) = 0$.

4. Solve the equations $\dfrac{\partial u}{\partial x} + \dfrac{\partial u}{\partial y} = 0$ and $\dfrac{\partial u}{\partial x} - \dfrac{\partial u}{\partial y} = 0$ by setting $u(x,y) = X(x) Y(y)$.

4. Uniform Convergence

The basic facts about uniformly convergent sequences and series are probably known to most readers. A quick review of the main theorems may, however, be helpful.

Definition 1. Let $D \subseteq \mathbb{R}$ be a nonempty set and, for each $n \in \mathbb{N}$, let $f_n \colon D \to \mathbb{K}$. We call $\{f_n\}$ a sequence of functions in D. We shall say that our sequence converges to the function $g \colon D \to \mathbb{K}$ uniformly over D if, given any $\varepsilon > 0$, there is a natural number N such that

$$|g(x) - f_n(x)| < \varepsilon$$

for all $x \in D$, whenever $n \geq N$.

Our first result, though not terribly surprising, is useful.

Lemma 1. Let $D \subseteq \mathbb{R}$ be a nonempty set. A sequence of functions $\{f_n\}$ in D converges uniformly over this set if, and only if, to each $\varepsilon > 0$ there corresponds a natural number N such that

$$|f_n(x) - f_m(x)| < \varepsilon$$

for all x in D, whenever both m and n exceed N.

Proof. Suppose that our condition holds for the sequence $\{f_n\}$. Then for each $x \in D$, $\{f_n(x)\} \subseteq \mathbb{K}$ is a Cauchy sequence (Chapter Zero, Section 5, Definition 5). Any such sequence converges to a point of $\mathbb{K}$ that we shall denote by $g(x)$. Now given $\varepsilon > 0$ we can find a natural number N such that $|f_n(x) - f_m(x)| < \varepsilon$ for all $x \in D$, whenever both m and n exceed N. Letting $m \to \infty$ in this inequality gives us $|f_n(x) - g(x)| \leq \varepsilon$ for all $x \in D$, whenever $n \geq N$. But this says $\{f_n\}$ converges to g uniformly over D, and the proof is complete in this case.

Now suppose that the given sequence $\{f_n\}$ of functions in D converges to a function g uniformly over this set. Then, given $\varepsilon > 0$, we can find a natural number N such that $|g(x) - f_n(x)| < \varepsilon/2$ for all x in D, whenever $n \geq N$. But then $|f_n(x) - f_m(x)| \leq |f_n(x) - g(x)| + |g(x) - f_m(x)| < \varepsilon$ for all x in D, whenever both m and n are $\geq N$. Thus $\{f_n\}$ satisfies our condition.

Given a sequence $\{u_j\}$ in D we may construct another sequence in this set by letting

$$f_1(x) = u_1(x)$$
$$f_2(x) = u_1(x) + u_2(x)$$
$$\cdots \cdots \cdots \cdots \cdots$$
$$f_n(x) = \sum_{j=1}^{n} u_j(x)$$
$$\cdots \cdots \cdots \cdots \cdots$$

When $\{f_n\}$ converges to a function u_0 uniformly over D we say that the series

$$\sum_{j=1}^{\infty} u_j(x)$$

is uniformly convergent to $u_0(x)$ over D. We sometimes write $u_0(x) = \sum_{j=1}^{\infty} u_j(x)$.

Lemma 1 has an important consequence:

Corollary 1. (Weierstrass M-test) Let $\{u_j\}$ be a sequence of functions in D and let $\{M_j\}$ be a sequence of positive numbers. Suppose that

(i) $|u_j(x)| \leq M_j$ for all $x \in D$, and each fixed j;

(ii) $\sum_{j=1}^{\infty} M_j$ is convergent.

Then the series $\sum_{j=1}^{\infty} u_j(x)$ converges uniformly over D.

Proof. Because ΣM_j is convergent we can, given $\varepsilon > 0$, choose a natural number N such that

$$\left|\sum_{j=1}^{m} M_j - \sum_{j=1}^{n} M_j\right| < \varepsilon$$

whenever both m and n exceed N. Thus if $m > n$,

$$\sum_{j=n+1}^{m} M_j < \varepsilon$$

whenever $n \geq N$. Now, for each n, let $f_n(x) = \sum_{j=1}^{n} u_j(x)$.

Then, when $m > n$,

$$|f_m(x) - f_n(x)| = |\sum_{j=1}^{m} u_j(x) - \sum_{j=1}^{n} u_j(x)|$$

$$= |\sum_{j=n+1}^{m} u_j(x)| \leq \sum_{j=n+1}^{m} |u_j(x)|$$

$$\leq \sum_{j=n+1}^{m} M_j$$

by (i). Combining this with our first observation we see that

$$|f_n(x) - f_m(x)| < \varepsilon \text{ when } m > n \geq N.$$

By Lemma 1 the sequence $\{f_n\}$, and hence the series $\Sigma u_j(x)$, converges uniformly over D.

There are theorems that tell us when a certain property that is shared by each term of a convergent sequence of functions is also shared by the limit of this sequence. The main applications of these results, for us, are to series of functions that are defined on some interval of $\mathbb{R}$ and converge uniformly over this interval. So let $[a,b] \subseteq \mathbb{R}$, let $u_j \colon [a,b] \to \mathbb{K}$ for each $j = 1, 2, \ldots$, and, for each n, let $f_n(x) = \sum_{j=1}^{n} u_j(x)$. We shall assume that $\{f_n\}$ converges to $u_0(x)$ uniformly over $[a,b]$.

Lemma 2. If $x_0 \in [a,b]$ and each $u_j(x)$ is continuous at this point, then u_0 is also continuous at this point. In particular, if each u_j is continuous on $[a,b]$, then u_0 is continuous on this interval.

Proof. First suppose that $x_0 \in (a,b)$. We may write

$$|u_0(x) - u_0(x_0)| \leq |u_0(x) - f_n(x)| + |f_n(x) - f_n(x_0)| + |f_n(x_0) - u_0(x_0)|$$

Now suppose that $\varepsilon > 0$ is given. We first choose, and fix, f_n so that

$$|u_0(x) - f_n(x)| < \frac{\varepsilon}{3}$$

for all x in $[a,b]$. For this n we use the continuity of f_n at x_0 to choose $\delta > 0$ such that
$$|f_n(x) - f_n(x_0)| < \frac{\varepsilon}{3}$$
whenever $|x - x_0| < \delta$. It then follows that
$$|u_0(x) - u_0(x_0)| < \varepsilon$$
whenever $|x - x_0| < \delta$, which shows that u_0 is continuous at x_0.

The cases $x_0 = a$ and $x_0 = b$ are left to the reader.

Lemma 3. If each of the functions $u_j(x)$ is continuous on $[a,b]$, then
$$\int_a^b u_0(x)dx = \int_a^b \left(\sum_{j=1}^\infty u_j(x)\right) dx = \sum_{j=1}^\infty \int_a^b u_j(x)dx$$

Proof. By lemma 2 the function $u_0(x)$ is continuous on $[a,b]$, and hence its Riemann integral exists. Now
$$|\int_a^b f_n(x)dx - \int_a^b u_0(x)dx| \leq \int_a^b |f_n(x) - u_0(x)|dx$$
and because $\{f_n\}$ converges to u_0 uniformly over $[a,b]$ we can, given $\varepsilon > 0$, choose a natural number N so that
$$|f_n(x) - u_0(x)| < \frac{\varepsilon}{b-a}$$
for all $x \in [a,b]$, whenever n exceeds N. Thus
$$|\int_a^b f_n(x)dx - \int_a^b u_0(x)dx| = |\sum_{j=1}^n \int_a^b u_j(x)dx - \int_a^b u_0(x)dx| < \varepsilon$$
whenever $n \geq N$. This proves the lemma.

One cannot, in general, find the derivative of $u_0(x)$ by differentiating the series $\Sigma u_j(x)$ term-by-term. In fact, $u_0(x)$ may not have a derivative even when every $u_j(x)$ does (see below). However, we do have the following:

Lemma 4. Suppose that each of the functions $u_j(x)$ is continuous on $[a,b]$, and has a continuous derivative at every point of (a,b). Suppose further that the series
$$\sum_{j=1}^\infty u_j'(x)$$

converges to a function $v(x)$ uniformly over (a, b). Then $u_0(x)$ is differentiable on this (open) interval and

$$u_0'(x) = v(x)$$

at each $x \in (a, b)$.

Proof. By our assumptions each of the functions

$$f_n(x) = \sum_{j=1}^{n} u_j(x)$$

is continuous on $[a, b]$ and has a continuous derivative on (a, b). Moreover, the sequence $\{f_n'(x)\}$ converges to $v(x)$ uniformly over (a, b). It follows, from lemma 2, that $v(x)$ is continuous on this interval.

Now given $x \in (a, b)$ we can choose a number $x_0 < x$ such that $[x_0, x] \subseteq (a, b)$. Then by lemma 3

$$\int_{x_0}^{x} v(t)dt = \lim_{n \to \infty} \int_{x_0}^{x} f_n'(t)dt = \lim_{n \to \infty} \{f_n(x) - f_n(x_0)\} = u_0(x) - u_0(x_0)$$

Thus

$$u_0(x) = \int_{x_0}^{x} v(t)dt + u_0(x_0)$$

It follows that u_0 is differentiable at x and that

$$u_0' = v(x)$$

A uniformly convergent series of even such well-behaved functions as sines and cosines can converge to a function with very strange properties. The following examples illustrate this point.

It is said that in a lecture given in 1861, Riemann asserted that the function

$$R(x) = \sum_{n=1}^{\infty} \frac{\sin n^2 \pi x}{n^2}$$

is continuous at every x but does not have a derivative at any point. The M-test shows that the series converges uniformly over $\mathbb{R}$ and so the limit function is certainly continuous by lemma 2. When Karl Weierstrass heard about Riemann's claim, he tried to prove that $R(x)$ was nowhere differentiable. He was unable to do so but, in the course of trying he came up with the following set of functions,

$$W_{a,b}(x) = \sum_{n=1}^{\infty} a^n \cos b^n \pi x$$

and was able to show that when $0 < a < 1$, b is an odd integer and $ab > 1 + \frac{3}{2}\pi$, these functions are nowhere differentiable. Again the M-test shows that the series converges uniformly over $\mathbb{R}$ when $0 < a < 1$; hence, the associated functions are continuous. In 1916, G. H. Hardy improved on this result by showing that the functions $W_{a,b}(x)$ are nowhere differentiable when $0 < a < 1$, $b > 1$, and $ab \geq 1$. Hardy also showed that $R(x)$ has no derivative when x is an irrational number or a rational number of a certain kind. Surprisingly, J. Gerver proved that $R(x)$ does have a derivative at certain other rational points. This he did in 1969 while he was an undergraduate student at Columbia University.

Exercises 4

For problems 1 and 2, T is the unit circle and G is the set of all points that are inside T.

* 1. Suppose that we have a sequence of functions each of which is continuous on $G \cup T$ and harmonic in G. If this sequence converges uniformly over T, show that it converges uniformly over $G \cup T$.

* 2. (a) For any fixed n show that $r^n (a_n \cos n\theta + b_n \sin n\theta)$ is harmonic in G.
    (b) For ρ fixed, $0 < \rho < 1$ show that $\sum \rho^n$, $\sum n\rho^n$, and $\sum n^2 \rho^n$ are convergent.
    (c) Suppose that, for some M, $|a_n| \leq M$ and $|b_n| \leq M$ all n. Set
    $$u(r, \theta) = \sum_{n=1}^{\infty} r^n (a_n \cos n\theta + b_n \sin n\theta)$$
    for all $(r, \theta) \in G$. Show that $u(r, \theta)$ is harmonic in G.

3. Recall that, for a given sequence $a_0, a_1, \ldots$ of real numbers, $\sum_{n=0}^{\infty} a_n x^n$ is called a(real) power series. To any series there corresponds an R, $0 \leq R \leq \infty$, such that $\sum a_n x^n$ converges for all x with $|x| < R$ and diverges for all x with $|x| > R$. R is called the radius of the series.

    (a) For each fixed ρ, $\rho < R$, shows that the series converges uniformly on $[-\rho, \rho]$.
    (b) For each $x \in (-R, R)$, define $f(x)$ to be $\sum_{n=0}^{\infty} a_n x^n$, and show that f is continuous on $(-R, R)$.
    (c) If $[a, b] \leq (-R, R)$ show that
    $$\int_a^b f(x) dx = \int_a^b \left(\sum_{n=0}^{\infty} a_n x^n \right) dx = \sum_{n=0}^{\infty} a_n \int_a^b x^n dx$$

(d) Show that $\sum_{n=0}^{\infty} x^n$ converges to $\frac{1}{1-x}$ at each point of $(-1, 1)$, but that convergence is *not* uniform over this interval.

4. Suppose that $\sum_{n=0}^{\infty} a_n x^n$ has radius R, $R \neq 0$ and suppose that the series $\sum_{n=1}^{\infty} n a_n x^{n-1}$ has radius R'.

(a) If $|x| < R$, choose x_0 so that $|x| < |x_0| < R$ and show that for some constant, call it A, we have $|a_n x_0^n| \leq A$ for all n.

(b) With x and x_0 as in (a) write
$$n a_n x^{n-1} = \frac{n}{x_0} a_n x_0^n \left(\frac{x}{x_0}\right)^{n-1}$$
and conclude that
$$|n a_n x^{n-1}| \leq \frac{A}{|x_0|} n r^{n-1}$$
where A is the number found in (a) and $r = \left|\frac{x}{x_0}\right|$.

(c) Show that $\sum_{n=1}^{\infty} \frac{A}{|x_0|} n r^{n-1}$ is convergent and conclude from this that $\sum n a_n x^{n-1}$ converges whenever $|x| < R$. Thus $R \leq R'$.

(d) Now suppose that $R < |x| < R'$ and note that
$$|a_n x^n| = |n a_n x^{n-1}| \left|\frac{x}{n}\right| < |n a_n x^{n-1}|$$
when $|x| < n$. Deduce a contradiction and conclude that $R' = R$.

(e) Show that $\sum_{n=0}^{\infty} a_n x^n$ is differentiable at every $x \in (-R, R)$ and that its derivative is $\sum_{n=1}^{\infty} n a_n x^{n-1}$.

5. Suppose that we have two power series, $\sum_{n=0}^{\infty} a_n x^n$ and $\sum_{n=0}^{\infty} b_n x^n$. Suppose that for some $r > 0$ we have $\sum a_n x^n = \sum b_n x^n$ whenever $|x| < r$. Show that $a_n = b_n$ for $n = 0, 1, 2, \ldots$.

5. The Formulas of Euler

The boundary data for the DP is a real-valued function that is continuous on the unit circle, on T. Let us denote the set of all such functions by $C_r(T)$. So every element of $C_r(T)$ is a real-valued function that is continuous on T or, equivalently, a real-valued function that is defined on $\mathbb{R}$ is continuous and is 2π-periodic. Our discussion thus far has led us to ask if every $f \in C_r(T)$ can be "represented" by a trigonometric series. In 1777 Euler made the first real progress toward answering this question. We shall present his results here.

Let us suppose that the trigonometric series

$$\frac{1}{2}a_0 + \sum_{n=1}^{\infty}(a_n \cos n\theta + b_n \sin n\theta) \tag{1}$$

converges to the function $f(\theta)$ uniformly over $\mathbb{R}$ (or, equivalently, over T). We are supposing also that the coefficients in this series are real, so clearly $f \in C_r(T)$. Now given $\varepsilon > 0$, we can find a natural number N such that

$$|f(\theta) - \{\frac{1}{2}a_0 + \sum_{n=1}^{m}(a_n \cos n\theta + b_n \sin n\theta)\}| < \varepsilon \tag{2}$$

for all θ, whenever $m \geq N$. We may write, for each n,

$$\cos n\theta = \frac{e^{in\theta} + e^{-in\theta}}{2}, \qquad \sin \theta = \frac{e^{in\theta} - e^{-in\theta}}{2i}$$

Putting these expressions into (2) we obtain

$$|f(\theta) - \{\frac{1}{2}a_0 + \sum_{n=1}^{m}\left(\frac{a_n - ib_n}{2}\right)e^{in\theta} + \sum_{n=1}^{m}\left(\frac{a_n + ib_n}{2}\right)e^{-in\theta}\}| < \varepsilon \tag{3}$$

or, setting $c_0 = \frac{1}{2}a_0$, $c_n = \frac{a_n - ib_n}{2}$, $c_{-n} = \frac{a_n + ib_n}{2}$ for each $n = 1, 2, \ldots$.

$$|f(\theta) - \sum_{n=-m}^{m} c_n e^{in\theta}| < \varepsilon \tag{4}$$

for all θ, when $m \geq N$.

Definition 1. When $D \subseteq \mathbb{R}$, $D \neq \emptyset$, and we have a set of functions $u_j : D \to \mathbb{K}$, one for each $j \in \mathbb{Z}$, we say that the series $\sum_{j=-\infty}^{\infty} u_j(x)$ converges to the function $v(x)$ uniformly over D whenever the sequence $f_0(x) = u_0(x)$, $f_1(x) = u_{-1}(x) + u_0(x) + u_1(x), \ldots, f_n(x) = \sum_{j=-n}^{n} u_j(x), \ldots$ converges to $v(x)$

uniformly over D. More precisely, we say that the series converges to $v(x)$ uniformly over D if, given $\varepsilon > 0$, we can find N such that

$$|v(x) - \sum_{j=-n}^{n} u_j(x)| < \varepsilon$$

for all $x \in D$, when $n \geq N$.

The inequalities (2) and (4) show that the series (1) converges to $f(\theta)$ uniformly over $\mathbb{R}$ if, and only if, the series

$$\sum_{-\infty}^{\infty} c_n e^{in\theta} \qquad (5)$$

converges to this function uniformly over $\mathbb{R}$.

We began by assuming that (1) converged to $f(\theta)$ uniformly over $\mathbb{R}$, and we have just seen that this is equivalent to assuming that (5) converges to $f(\theta)$ uniformly over $\mathbb{R}$. The results we want can be obtained, very easily, from the latter series.

Choose, and fix, an integer p. Then the series

$$\left(\sum_{-\infty}^{\infty} c_n e^{in\theta} \right) e^{-ip\theta} = \sum_{-\infty}^{\infty} c_n e^{i(n-p)\theta}$$

converges to $f(\theta)e^{-ip\theta}$ uniformly over $\mathbb{R}$; equivalently, uniformly over T. Thus by lemma 4.3 we have

$$\int_{-\pi}^{\pi} f(\theta) e^{-ip\theta} d\theta = \int_{-\pi}^{\pi} \left(\sum_{-\infty}^{\infty} c_n e^{i(n-p)\theta} \right) d\theta$$

$$= \sum_{-\infty}^{\infty} c_n \int_{-\pi}^{\pi} e^{i(n-p)\theta} d\theta$$

It is easy to evaluate these integrals. Clearly,

$$\int_{-\pi}^{\pi} e^{i(n-p)\theta} d\theta = 0$$

when $n \neq p$, and when $n = p$, the integral is equal to 2π.

Thus our last infinite series reduces to a single term: $2\pi c_p$. And this gives

$$c_p = \frac{1}{2\pi} \int_{-\pi}^{\pi} f(\theta) e^{-ip\theta} d\theta \qquad \text{all } p \in Z \qquad (6)$$

Thus the coefficients in the series (5) can be computed from $f(\theta)$.

The coefficients in the series (1) can also be computed from this function. We have $a_0 = 2c_0$, and a little algebra gives $a_n = c_n + c_{-n}$, $b_n = i(c_n - c_{-n})$, $n = 1, 2, \ldots$. Hence (6) gives us

$$a_n = \frac{1}{\pi} \int_{-\pi}^{\pi} f(\theta) \cos n\theta \, d\theta, \qquad n = 0, 1, 2, \ldots \tag{7}$$

$$b_n = \frac{1}{\pi} \int_{-\pi}^{\pi} f(\theta) \sin n\theta \, d\theta, \qquad n = 1, 2, \ldots \tag{8}$$

and here we see why it is customary to put the $\frac{1}{2}$ before the a_0. Without this factor the formula (7) would not be valid when $n = 0$.

Given a function $f \in C_r(T)$, the numbers obtained from f using formulas (7) and (8) are called the (real) Fourier coefficients of f. We sometimes denote them by $a_n(f)$, $b_n(f)$. The trigonometric series (1) whose coefficients are the Fourier coefficients of f is called the (real) Fourier series of f.

The numbers obtained from f by using (6) are called the complex Fourier coefficients of f and are sometimes written $c_n(f)$. The series (5) is called the complex Fourier series of f when its coefficients are the numbers $c_n(f)$.

We may now ask whether the Fourier series of a given function converges to that function. This is a tricky matter. The discussion of these series is intricate and technically challenging, and it is not even possible to state many of the results unless we first discuss the Lebesgue integral. One of the major reasons for interest in that integral is the profound impact it had on the study of Fourier series. In the next few sections we discuss the convergence of the Fourier series of a function in $C_r(T)$ and use the results we obtain to solve the DP. A discussion of these series for more general functions is given in the next chapter.

We end this section with an important theorem due to Riemann.

Theorem 1. For any $f \in C_r(T)$, $\lim_{|n| \to \infty} c_n(f) = 0$.

Proof. Given $f \in C_r(T)$ we may write

$$-c_n(f) = e^{i\pi} c_n(f) = \frac{1}{2\pi} \int_{-\pi}^{\pi} f(t) e^{-in(t - \frac{\pi}{n})} dt$$

In this integral we set $u = t - \frac{\pi}{n}$ to obtain

$$-c_n(f) = \frac{1}{2\pi} \int_{-\pi - \frac{\pi}{n}}^{\pi - \frac{\pi}{n}} f(u + \frac{\pi}{n}) e^{-inu} du = \frac{1}{2\pi} \int_{-\pi}^{\pi} f(u + \frac{\pi}{n}) e^{-inu} du$$

where we have used problem 3 of Exercises 2. Now

$$2c_n(f) = c_n(f) - [-c_n(f)]$$

and so we have
$$2c_n(f) = \frac{1}{2\pi}\int_{-\pi}^{\pi}\{f(t) - f(t+\frac{\pi}{n})\}e^{-int}dt$$

From this we find that
$$2|c_n(f)| \leq \frac{1}{2\pi}\int_{-\pi}^{\pi}|f(t) - f(t+\frac{\pi}{n})||e^{-int}|dt \leq \max_{-\pi \leq t \leq \pi}|f(t) - f(t+\frac{\pi}{n})|$$

Now $f \in C(T)$, and so this function is uniformly continuous on $\mathbb{R}$ (Exercises 2, problem 7). Thus, given $\epsilon > 0$ we can choose a natural number N such that
$$|f(t) - f\left(t + \frac{\pi}{n}\right)| < 2\varepsilon$$
whenever $|n| \geq N$, and this holds for any $t \in [-\pi, \pi]$. It follows that
$$|c_n(f)| \leq \varepsilon$$
whenever $|n| \geq N$, and this proves our theorem.

Corollary 1. For any function $f \in C_r(T)$ we have $\lim_{n \to \infty} a_n(f) = 0$ and $\lim_{n \to \infty} b_n(f) = 0$.

Proof. Because $a_n(f) = c_n(f) + c_{-n}(f)$ and $b_n(f) = i\{c_n(f) - c_{-n}(f)\}$ the theorem shows that $\lim_{n \to \infty} a_n(f) = 0 = \lim_{n \to \infty} b_n(f)$.

Exercises 5

1. Define $f(x) = x^2$ for $-\pi \leq x \leq \pi$, and $f(x + 2\pi) = f(x)$ for all x. Clearly, $f \in C_r(T)$.

   (a) Compute the real and the complex Fourier coefficients of f and show that the real and complex Fourier series of this function converge uniformly over $\mathbb{R}$.

   (b) *Assuming* that the real Fourier series of f converges to this function, compute the numbers (i) $\sum_{n=1}^{\infty} \frac{1}{n^2}$ and (ii) $\sum_{n=1}^{\infty} \frac{(-1)^{n+1}}{n^2}$.

*2. For $f \in C_r(T)$ show that the Fourier coefficients of f are bounded; i.e., show that there is a real number μ' such that $|a_n(f)| \leq \mu'$ and $|b_n(f)| \leq \mu'$ for all n.

3. If $f \in C_r(T)$ has Fourier coefficients $\{c_n\}$ and if f' exists and is also in $C_r(T)$, show that $\lim_{|n| \to \infty} nc_n = 0$. If f'' also exists and is in $C_r(T)$, show that $\lim_{|n| \to \infty} n^2 c_n = 0$.

6. Cesàro Convergence

A modified form of convergence was introduced into mathematics by the Italian mathematician Ernesto Cesàro. This notion seems, at first glance at least, somewhat contrived and, perhaps, artificial. A major result of Leopold Fejér (next section) shows, however, that this type of convergence is the natural one to use in connection with the Fourier series of a function in $C_r(T)$.

Definition 1. Given a sequence a_0, a_1, a_2, ... in $\mathbb{K}$ the numbers $\sigma_1 = a_0$, $\sigma_2 = \frac{a_0+a_1}{2}$, $\sigma_3 = \frac{a_0+a_1+a_2}{3}$, ... are called the Cesàro means of this sequence. We shall say that $\{a_n\}_{n=0}^{\infty}$ is Cesàro convergent to the number σ if $\lim_{n\to\infty} \sigma_n = \sigma$.

One can easily show that a convergent sequence is Cesàro convergent to its (ordinary) limit. There are, however, sequences that do not converge in the ordinary sense that are convergent in this new sense (see Exercises 6, problems 1 and 2). So Cesàro convergence "contains" and extends the ordinary notion of convergence.

As one might expect, a series is said to be Cesàro convergent (or, sometimes, Cesàro summable) to the number s' if its sequence of partial sums is Cesàro convergent to s'. This can get a bit messy. The mth Cesàro mean, call it σ_m, for the series $\sum_{-\infty}^{\infty} a_k$ is

$$\sigma_m = \frac{\sum_{k=0}^{0} a_k + \sum_{k=-1}^{1} a_k + \sum_{k=-2}^{2} a_k + \cdots + \sum_{k=-(m-1)}^{m-1} a_k}{m} \quad (1)$$

We shall now derive an expression for the mth Cesàro mean of the complex Fourier series of a function $f \in C_r(T)$. It turns out that there is a nice, compact, expression for this mean in terms of an integral.

Given $f \in C_r(T)$ the nth partial sum of its complex Fourier series is

$$s_n(x;f) = \sum_{k=-n}^{n} c_k(f)e^{ikx} = \sum_{k=-n}^{n} e^{ikx} \left[\frac{1}{2\pi}\int_{-\pi}^{\pi} f(y)e^{-iky}dy\right] \quad (2)$$

$$= \frac{1}{2\pi}\int_{-\pi}^{\pi} f(y)\left[\sum_{k=-n}^{n} e^{ik(x-y)}\right] dy$$

The mth Cesàro mean of this series is

$$\sigma_m(x;f) = \frac{s_0(x;f) + s_1(x;f) + \cdots + s_{m-1}(x;f)}{m} \quad (3)$$

Combining this with (2) gives us

$$\sigma_m(x;f) = \frac{1}{2\pi}\int_{-\pi}^{\pi} f(y) \frac{\left[\sum_{k=0}^{0} e^{ik(x-y)} + \cdots + \sum_{k=-(m-1)}^{m-1} e^{ik(x-y)}\right]}{m} dy \quad (4)$$

Observe that the integrand contains the mth Cesàro mean of the series

$$\sum_{k=-\infty}^{\infty} e^{ik(x-y)} \tag{5}$$

If we denote this mean by $K_m(x-y)$, then (4) becomes

$$\sigma_m(x; f) = \frac{1}{2\pi} \int_{-\pi}^{\pi} f(y) K_m(x-y) dy, \quad m = 1, 2, 3 \cdots \tag{6}$$

We shall use this expression in the next section.

The functions $K_m(x)$, $m = 1, 2, \ldots$, are the Cesàro means of the series $\sum_{-\infty}^{\infty} e^{ikx}$. These functions have some useful properties:

Lemma 1. For each fixed m we have

(i) $K_m(x) = \frac{1}{m}\left(\frac{1-\cos mx}{1-\cos x}\right) = \frac{1}{m}\left(\frac{\sin \frac{1}{2}mx}{\sin \frac{1}{2}x}\right)^2$

(ii) $K_m(x) \geq 0$ for all x

(iii) $\frac{1}{2\pi} \int_{-\pi}^{\pi} K_m(x) dx = 1$

Furthermore, if I is any open interval containing zero, then

$$\lim_{m \to \infty} \sup\{K_m(x) \mid x \in (-\pi, \pi] \setminus I\} = 0$$

Proof. Let us set $s_n(x) = \sum_{-n}^{n} e^{ikx}$ so that we may write

$$K_{n+1}(x) = \frac{s_n(x) + s_{n-1}(x) + \cdots + s_1(x) + s_0(x)}{n+1}$$

and

$$K_n(x) = \frac{s_{n-1}(x) + s_{n-2}(x) + \cdots + s_1(x) + s_0(x)}{n}$$

then

$$(n+1)K_{n+1}(x) - nK_n(x) = s_n(x) = \sum_{-n}^{n} e^{ikx} = \sum_{0}^{n} e^{ikx} + \sum_{1}^{n} e^{-ikx} \tag{1}$$

These last two series are (finite) geometric series and can be summed to give

$$\frac{1 - e^{i(n+1)x}}{1 - e^{ix}}$$

and
$$\frac{1 - e^{-i(n+1)x}}{1 - e^{-ix}} - 1$$
respectively. Adding these expressions and doing a little algebra gives us
$$(n+1)K_{n+1}(x) - nK_n(x) = \frac{\cos nx - \cos(n+1)x}{1 - \cos x} \tag{2}$$
Now $K_1(x) = 1$ for all x, and putting this into (2) we find $K_2(x)$. Putting $K_2(x)$ into (2) gives us $K_3(x)$, and so on. In this way we find that
$$K_m(x) = \frac{1}{m}\left(\frac{1 - \cos mx}{1 - \cos x}\right)$$
which is the first expression in (i). Using our knowledge of the trigonometric functions we can transform this into
$$\frac{1}{m}\left(\frac{\sin \frac{1}{2}mx}{\sin \frac{1}{2}x}\right)^2$$
which is the second expression in (i). Note that this proves (ii) as well.

To prove (iii) we consider the function that is 1 for all x. The Fourier coefficients of this function are
$$c_k = \frac{1}{2\pi}\int_{-\pi}^{\pi} 1 e^{-ikx} dx$$
for $k \in Z$. Clearly, $c_k = 0$ when $k \neq 0$, and $c_0 = 1$. Thus the Fourier series of this function consists of a single term, which is 1. Now when we compute the mth Cesàro mean of this series, the 1 occurs m times in the numerator. But we also divide by m. Hence
$$\sigma_m(x; f \equiv 1) = 1$$
for all x and each $m = 1, 2, \ldots$. We now refer to equation (6) above. For the function we are treating here this equation becomes
$$1 = \frac{1}{2\pi}\int_{-\pi}^{\pi} 1 \cdot K_m(x - y) dy$$
This last expression is true for any x, so we may take $x = 0$. Finally, because $K_m(-y) = K_m(y)$ by (i), we have proved (iii).

Now suppose that $0 < \delta < \pi$ and $\delta \leq |x| \leq \pi$. Then $\frac{1}{2}x$ and $\frac{1}{2}\delta$ are both in the interval where the sine function is monotonically increasing. Hence
$$\left(\sin \frac{1}{2}\delta\right)^2 \leq \left(\sin \frac{1}{2}x\right)^2$$

Given an open interval I that contains zero, we may choose $\delta > 0$ so that $(-\delta, \delta) \subseteq I$. Then

$$\sup\{K_n(x) \mid x \in (-\pi, \pi] \setminus I\} \leq \sup\{K_n(x) \mid \delta \leq |x| \leq \pi\}$$

But this last term is less than or equal to

$$\sup \frac{1}{n}\left(\frac{\sin \frac{1}{2}nx}{\sin \frac{1}{2}x}\right)^2 \leq \frac{1}{n|\sin \frac{1}{2}\delta|^2}$$

Letting $n \to \infty$ in this last expression proves our claim.

Exercises 6

1. Show that the sequence $\{(-1)^{n+1}\}_{n=1}^\infty$, which is clearly divergent, is Cesàro convergent zero.

* 2. If the sequence $\{a_n\}_{n=1}^\infty$ is convergent to b, show that it is also Cesàro convergent to b.

* 3. We call $\{k_m\}_{n=1}^\infty$ Fejér's kernal. Show that $K_m(x) = \sum_{|k|<m}(1 - \frac{|k|}{m})e^{-ikx}$.

4. Suppose that the series $\sum_{n=1}^\infty a_n$ is Cesàro convergent to b. Suppose further that the sequence $\{na_n\}_{n=1}^\infty$ is bounded. Show that the original series must converge to b. Results of this kind (i.e., results that give conditions under which a series that converges in some general sense must converge in the ordinary sense) are called Tauberian theorems.

5. Let $\{a_n\}$ and $\{b_n\}$ be two sequences of real numbers. Suppose that
    (a) i. $a_n \geq a_{n+1}$ for all n;
         ii. $\lim_{n \to \infty} a_n = 0$;
         iii. The series $\sum b_n$ has bounded partial sums.
    (b) Show that $\sum_n a_n b_n$ is convergent. *Hint:* Let $s_0 = b_0$, $s_1 = b_0 + b_1, \ldots, s_n = \sum_{j=0}^n b_j$. Write $a_0 b_0 + a_1 b_1 + \cdots$ as $a_0 s_0 + a_1(s_1 - s_0) + a_2(s_2 - s_1) + \cdots$.
    (c) Show that $\sum a_n b_n$ is convergent even if the b_n's are complex, just as long as the partial sums of $\sum b_n$ are bounded.

(d) Show that $\sum_{n=1}^{\infty} \dfrac{i^n}{n}$ is convergent. More generally, show that $\sum_{n=1}^{\infty} \dfrac{z^n}{n}$ converges when $|z| = 1$ and $z \neq 1$.

(e) If we take $b_n = (-1)^n$, show that this test, which is due to the Norwegian mathematician Niels Abel, gives us the familiar Leibnitz test for alternating series.

7. Fejér's Theorem

Scientists who are engaged in SETI (the search for extraterrestrial intelligence) call it the Fermi paradox. It first arose during a lunchtime conversation among the physicists Enrico Fermi, Emil Konopinski, Edward Teller, and Herbert York, which took place at Los Alamos sometime in the year 1950. They were speculating about when we could expect to achieve faster-than-light travel (assuming this is possible). Each man gave his opinion, and the talk turned for awhile to scientific subjects of more immediate interest, when Fermi suddenly asked "So where is everybody?" It was an abrupt change of subject, but everyone knew what he meant. Given the age of the universe and the widely held belief that technically sophisticated civilizations are fairly common, even the most pessimistic estimate of when faster-than-light travel is achieved implied that visits from "elsewhere" should be, not frequent perhaps, but not really uncommon either. Yet we have had none! This question has been discussed by scientists ever since, but with no general agreement on an answer. It is said that when the Hungarian physicist Leo Szilard heard of it he quipped, "They are among us, but they call themselves Hungarians."

Szilard was famous for his off-beat sense of humor and, no doubt, would claim that his facetious remark also explained why the tiny country of Hungary has produced so many outstanding mathematicians. We shall discuss the work of one of them, Leopold Fejér, here. We shall then use his result to complete our solution to the DP.

Theorem 1. The Cesàro means of the Fourier series of any real-valued function that is continuous on the unit circle converge to that function uniformly over the circle.

Proof. Let $f \in C_r(T)$ and let $\sigma_n(x)$ denote the nth Cesàro mean of its (complex) Fourier series. Given $\varepsilon > 0$, we shall show that there is a natural number N such that

$$|\sigma_n(x) - f(x)| < \varepsilon$$

for all $x \in T$, whenever $n \geq N$. We shall use formula (6) and Lemma 1 of Section 6. From (iii) of that lemma we may write, for any fixed x,

$$\sigma_n(x) - f(x) = \frac{1}{2\pi} \int_{-\pi}^{\pi} f(t) K_n(x-t) dt - \frac{1}{2\pi} \int_{-\pi}^{\pi} K_n(t) f(x) dt$$

In the first of these integrals we set $u = x - t$ and use the 2π-periodicity of the integrand to write

$$\int_{-\pi}^{\pi} f(t) K_n(x-t) dt = -\int_{x+\pi}^{x-\pi} f(x-u) K_n(u) du$$
$$= -\int_{\pi}^{-\pi} f(x-u) K_n(u) du = \int_{-\pi}^{\pi} f(x-u) K_n(u) du$$

We may now write

$$\sigma_n(x) - f(x) = \frac{1}{2\pi} \int_{-\pi}^{\pi} \{f(x-t) - f(x)\} K_n(t) dt$$

From this last expression we obtain

$$|\sigma_n(x) - f(x)| \leq \frac{1}{2\pi} \int_{-\pi}^{\pi} |f(x-t) - f(x)| K_n(t) dt \qquad (1)$$

$$= \frac{1}{2\pi} \int_{-\delta}^{\delta} |f(x-t) - f(x)| K_n(t) dt + \frac{1}{2\pi} \int_{|t| \geq \delta} |f(x-t) - f(x)| K_n(t) dt$$

where $\delta > 0$ is to be determined.

The first of these integrals can be estimated as follows:

$$\frac{1}{2\pi} \int_{-\delta}^{\delta} |f(x-t) - f(x)| K_n(t) dt \qquad (2)$$
$$\leq \sup_{-\delta \leq t \leq \delta} |f(x-t) - f(x)| \frac{1}{2\pi} \int_{-\delta}^{\delta} K_n(t) dt$$
$$\leq \sup_{-\delta \leq t \leq \delta} |f(x-t) - f(x)| \frac{1}{2\pi} \int_{-\pi}^{\pi} K_n(t) dt = \sup_{-\delta \leq t \leq \delta} |f(x-t) - f(x)|$$

where our change in the limits of integration is justified by (ii) of Lemma 1 of Section 6 and we have also used (iii) of the lemma.

Let us now estimate the second integral appearing in (1). We have

$$\frac{1}{2\pi} \int_{|t| \geq \delta} |f(x-t) - f(x)| K_n(t) dt \qquad (3)$$
$$\leq \frac{2}{2\pi} \left[\max_{-\pi \leq x \leq \pi} |f(x)| \right] \{ \sup_{|t| \geq \delta} K_n(t) \} 2\pi$$

Now suppose $\varepsilon > 0$ is given. We use the uniform continuity of f to find $\delta > 0$ such that

$$|f(x-t) - f(x)| < \frac{\varepsilon}{2}$$

whenever $t \in [-\delta, \delta]$ and $x \in [-\pi, \pi]$. With δ now fixed we see by (2) that the first integral on the right-hand side of (1) is less than $\frac{\varepsilon}{2}$.

To finish the proof, we use the last statement in Lemma 1 of Section 6. It tells us that we can choose a natural number N so that

$$\sup_{|t| \geq \delta} K_n(t) < \frac{\varepsilon}{4 \max\limits_{-\pi \leq t \leq \pi} |f(t)|}$$

whenever $n \geq N$. Combining this with (3) we see that the second integral on the right-hand side of (1) is less than $\frac{\varepsilon}{2}$ whenever $n \geq N$.

Definition 1. A sequence $\{f_n\}$ of functions defined on a set D is said to converge pointwise on D if for each $x \in D$ the sequence of numbers $\{f_n(x)\}$ is convergent. A series $\Sigma u_j(x)$ of functions defined in D is pointwise convergent in D if its sequence of partial sums converges pointwise on D.

If $f \in C_r(T)$ and if the Fourier series of f converges at $\theta_0 \in T$, then its limit must be $f(\theta_0)$. This follows from Fejér's theorem. The partial sums of the Fourier series, when $\theta = \theta_0$, converge and so their Cesàro means must converge to the (ordinary) limit. But we know that these means converge to $f(\theta_0)$, and this proves our claim. We have shown the following:

Corollary 1. If $f \in C_r(T)$ and if the Fourier series of f converges at any point $\theta_0 \in T$, then this series converges to $f(\theta_0)$.

Remark 1. We have seen that the Fourier series for the function $f(x) = x^2$, $-\pi \leq x \leq \pi$, $f(x + 2\pi) = f(x)$ all x, converges uniformly over T. Because uniform convergence over T clearly implies pointwise convergence on T, we now know that the limit of this series is $f(x)$. This establishes the results found in Exercises 5, problem 1.

Let $n_1, n_2, \ldots, n_\ell$ be a finite set of integers, some of which may be negative, and let $c_1, c_2, \ldots, c_\ell$ be complex numbers. The function

$$\tau(\theta) = \sum_{j=1}^{\ell} c_j e^{in_j \theta}$$

is called a trigonometric polynomial. It is clear that $\tau(\theta)$ is in $C(T)$. Moreover, the Cesàro means of any $f \in C_r(T)$ are trigonometric polynomials. Thus:

Corollary 2. Given any $f \in C_r(T)$ and any $\varepsilon > 0$ there is a trigonometric polynomial $\tau_\varepsilon(\theta)$ such that

$$|f(\theta) - \tau_\varepsilon(\theta)| < \varepsilon$$

for all $\theta \in T$.

Corollary 3. (Weierstrass Approximation Theorem) A continuous real-valued function defined on a closed bounded interval can be approximated, as closely as we wish, on that interval by a polynomial.

Proof. Let a, b be real numbers with $a < b$, and let f be a real-valued function that is defined and continuous on $[a, b]$. Given $\varepsilon > 0$ we shall construct a polynomial $P_\varepsilon(x)$ such that $|f(x) - P_\varepsilon(x)| < \varepsilon$ for all $x \in [a, b]$.

Let us suppose, first, that $[a, b] \subseteq (-\pi, \pi)$. Then we may extend f, in any convenient way, to a function $\tilde{f}$ that is continuous on $[-\pi, \pi]$ and satisfies $\tilde{f}(-\pi) = \tilde{f}(\pi)$. Then, by Corollary 2, we can find a trigonometric polynomial $\tau(x)$ such that $|\tilde{f}(x) - \tau(x)| < \frac{\varepsilon}{2}$ for all $x \in [-\pi, \pi]$. Now $\tau(x)$ can be written as $\sum_{j=1}^{\ell} c_j e^{in_j x}$, where the c_j's are complex numbers, the n_j's are integers, and ℓ is a natural number. For each fixed j we may expand the function $e^{in_j x}$ in a MacLaurin series that converges to this function uniformly over $[-\pi, \pi]$. Hence for each $j = 1, 2, \ldots, \ell$ we can choose a natural number k_j such that $\max_{-\pi \leq x \leq \pi} |c_j e^{in_j x} - c_j \sum_{k=0}^{k_j} \frac{(in_j x)^k}{k!}| < \frac{\varepsilon}{2\ell}$. Thus the function $c_j e^{in_j x}$ can be approximated, to within $\frac{\varepsilon}{2\ell}$, by a polynomial $P_j(x) = c_j \sum_{k=0}^{k_j} \frac{(in_j x)^k}{k!}$ over $[-\pi, \pi]$. It follows that

$$\max_{-\pi \leq x \leq \pi} |\tilde{f}(x) - \sum_{j=1}^{\ell} P_j(x)| \leq$$

$$\max_{-\pi \leq x \leq \pi} |\tilde{f}(x) - \sum_{j=1}^{\ell} c_j e^{in_j x}| + \max_{-\pi \leq x \leq \pi} |\sum_{j=1}^{\ell} c_j e^{in_j x} - \sum_{j=1}^{\ell} P_j(x)| \leq$$

$$\frac{\varepsilon}{2} + \sum_{j=1}^{\ell} \max |c_j e^{in_j x} - P_j(x)| \leq \frac{\varepsilon}{2} + \frac{\varepsilon}{2\ell} \cdot \ell = \varepsilon$$

Now any finite sum of polynomials is a polynomial, so we may set $P_\varepsilon(x) = \sum_{j=1}^{\ell} P_j(x)$, and the proof is complete in this case.

If $[a, b]$ is not contained in $(-\pi, \pi)$ we first define $\varphi(x) = (b-a)x+a$ and note that this function is continuous and maps $[0, 1]$ onto $[a, b]$. The inverse (Chapter Zero, Exercises 2, problem 2b) of φ, $\varphi^{-1}(x) = \frac{x-a}{b-a}$, is also continuous and, of course, maps $[a, b]$ onto $[0, 1]$.

Now $f \circ \varphi$ is a continuous function on $[0, 1]$ and so by the first part of the proof we can, given $\varepsilon > 0$, find a polynomial $P(x)$ such that $\max_{0 \leq x \leq 1} |(f \circ \varphi - P)(x)| < \varepsilon$. It follows then that $\max_{a \leq x \leq b} |(f - P \circ \varphi^{-1})x| = \max_{a \leq x \leq b} |(f \circ \varphi - P) \circ \varphi^{-1}(x)| \leq \max_{0 \leq x \leq 1} |(f \circ \varphi - P)(x)| < \varepsilon$. Because $P \circ \varphi^{-1}$ is clearly a polynomial, we are done.

Remark 2. It is customary to define harmonic functions to be real-valued, and consequently the boundary data for the DP must also be real-valued. This is why we restricted our attention to functions in $C_r(T)$. It is clear, however, that if f is a continuous complex-valued function on T, we may use formula (6) of Section 5 to define numbers $c_n(f)$ and formulas (7) and (8) of that section to define numbers $a_n(f)$ and $b_n(f)$. These numbers are called the Fourier coefficients of f. The series $\sum c_n(f)e^{in\theta}$ is called the complex form of the Fourier series for f, and the series $\frac{1}{2}a_0(f) + \sum(a_n(f)\cos n\theta + b_n(f)\sin n\theta)$ is called the real form of the Fourier series for this function. This terminology, although a little misleading, should not cause any problems. Observe that from Exercises 2, problem 6c, we must have $\lim_{|n| \to \infty} c_n(f) = 0$, $\lim_{n \to \infty} a_n(f) = \lim_{n \to \infty} b_n(f) = 0$. Furthermore, for every $\epsilon > 0$, there is a trigonometric polynomial $\tau_\epsilon(\theta)$ such that $|f(\theta) - \tau_\epsilon(\theta)| < \epsilon$ for all $\theta \in T$.

Exercises 7

* 1. Given any $f \in C_r(T)$, show that there is a sequence of trigonometric polynomials τ_n that converges to f uniformly over T.

* 2. (a) Show that the complex Fourier series of any trigonometric polynomial $\tau(\theta)$ is $\tau(\theta)$.

  (b) If $\tau(\varphi) = \sum_{j=1}^{\ell} c_j e^{in_j \varphi}$, find the real Fourier series of $\tau(\varphi)$.

* 3. Show that $-1 + 2Re\left[\frac{1}{1-re^{i\omega}}\right]$ is equal to $\frac{1-r^2}{1+r^2-2r\cos\omega}$ (Chapter Zero, Exercises 4, problem 6).

4. Let $f(x) = |x|$ for $-\pi \le x \le \pi$ and let $f(x + 2\pi) = f(x)$ for all x. Then $f \in C_r(T)$.

   (a) Compute the real Fourier series of $f(x)$.

   (b) Discuss the convergence of this Fourier series and find its limit.

5. Suppose that $f \in C_r(T)$ and that f' and f'' both exist and are in $C_r(T)$. Show that the Fourier series for f converges to this function uniformly over T.

* 6. If $f \in C_r(T)$ and if every Fourier coefficient of f is zero, show that f is identically zero on T.

7. If f is a real-valued function that is defined and continuous on $[a, b]$, show that there is a sequence $\{P_n(x)\}$ of polynomials that converges to $f(x)$ uniformly over this interval.

8. Let f be a real-valued function that is defined and continuous on $[0, 1]$. The moments of f are the numbers $\int_0^1 f(x)x^n dx$, where $n = 0, 1, 2, \ldots$. If two functions have the same moments, show that they are identical.

9. If f is a continuous real-valued function on $[0, 1]$, we set $B_n(f, x) = \sum_{k=0}^{n} \binom{n}{k} f\left(\frac{k}{n}\right) x^k(1-x)^{n-k}$ for each $n = 1, 2, \ldots$. These are the Bernstein polynomials of f. It can be shown that they converge to f uniformly over $[0, 1]$.

8. At Last the Solution

Let us now solve the DP for the unit circle T with boundary data f. The real Fourier coefficients of f, $\{a_n\}$, $\{b_n\}$, are bounded (Exercises 5, problem 2) and so the function

$$u(r, \theta) = \frac{1}{2}a_0 + \sum_{n=1}^{\infty} r^n(a_n \cos n\theta + b_n \sin n\theta) \tag{1}$$

is harmonic in G (recall that G is the interior of T). This was shown in Exercise 4, problem 2. Using formulas (7) and (8) of Section 5 we may write

$$u(r,\theta) = \frac{a_0}{2} + \sum_{n=1}^{\infty} r^n(a_n \cos n\theta + b_n \sin n\theta) =$$

$$\frac{1}{2\pi}\int_0^{2\pi} f(\varphi)d\varphi + \frac{1}{\pi}\sum_{n=1}^{\infty} r^n \left\{ \left(\int_0^{2\pi} f(\varphi)\cos n\varphi d\varphi\right)\cos n\theta \right.$$

$$\left. + \left(\int_0^{2\pi} f(\varphi)\sin n\varphi d\varphi\right)\sin n\theta \right\}$$

$$= \frac{1}{2\pi}\int_0^{2\pi} f(\varphi)d\varphi + \frac{1}{\pi}\sum_{n=1}^{\infty}\int_0^{2\pi} f(\varphi)r^n \cos n(\varphi-\theta)d\varphi$$

The series (1) converges uniformly over any closed subdisk of G (i.e., when $r \leq r_0$ where $r_0 < 1$ is fixed), and so by Lemma 3 of Section 4 this last sum can be written as follows:

$$u(r,\theta) = \frac{1}{2\pi}\int_0^{2\pi} f(\varphi)\left\{1 + 2\sum_{n=1}^{\infty} r^n \cos n(\varphi-\theta)\right\}d\varphi$$

In the series inside the integral we set $\varphi - \theta = \omega$ and note that

$$1 + 2\sum_{n=1}^{\infty} r^n \cos n\omega = -1 + 2\sum_{n=0}^{\infty} r^n \cos n\omega =$$

$$-1 + 2\operatorname{Re}\left[\sum_{n=0}^{\infty} r^n e^{in\omega}\right] = -1 + 2\operatorname{Re}\left[\frac{1}{1-re^{i\omega}}\right]$$

$$= \frac{1-r^2}{1+r^2-2r\cos\omega}$$

where Re denotes the real part (Chapter Zero, Exercises 4, problem 6) of the expression in brackets and we have used problem 3 of Exercises 7. Thus

$$u(r,\theta) = \frac{1}{2\pi}\int_0^{2\pi} f(\varphi)\frac{1-r^2}{1+r^2-2r\cos(\varphi-\theta)}d\varphi \qquad (2)$$

and we have shown that when $f \in C_r(T)$, the function defined by this integral is harmonic in G.

We now want to do two things. First, we want to show that the series (1) solves the DP; i.e., we want to show that $u(1,\theta) = f(\theta)$ on T. Second, we want to show that (2) solves the DP in the sense that

$$f(\theta) = u(1,\theta) = \lim_{r\to 1^-} \frac{1}{2\pi}\int_0^{2\pi} f(\varphi)\frac{1-r^2}{1+r^2-2r\cos(\varphi-\theta)}d\varphi$$

This gives us a nice, compact, representation of the solution: We shall do this in two steps.

Suppose, temporarily, that the Fourier coefficients of f are such that the series

$$\frac{1}{2}|a_0| + \sum_{n=1}^{\infty}(|a_n| + |b_n|) \tag{3}$$

is convergent. Then the Fourier series for f converges uniformly over T, and so its limit must be $f(\theta)$ (Section 7, Corollary 1). Thus the series (1) converges uniformly over $G \cup T$ to a function $u(r, \theta)$ that is continuous on this set and, moreover, $u(1, \theta) = f(\theta)$. We have already seen that $u(r, \theta)$ is harmonic in G, and so under the assumption (3), the function $u(r, \theta)$ given by (1) is the solution to the DP. In particular, we must have

$$\lim_{r \to 1^-} u(r, \theta) = f(\theta)$$

and because in G, $u(r, \theta)$ is given by (2), we must also have

$$\lim_{r \to 1^-} \frac{1}{2\pi} \int_0^{2\pi} f(\varphi) \frac{1 - r^2}{1 + r^2 - 2r\cos(\varphi - \theta)} d\varphi = f(\theta)$$

So we have proved the two things we wanted to prove, but only for functions $f \in CT(T)$ that satisfy (3).

Now suppose that our boundary data are merely continuous on T. We may choose a sequence of trigonometric polynomials $\{\tau_n(\theta)\}$ that converges to $f(\theta)$ uniformly over T (Section 7, Corollary 2). Now each $\tau_n(\theta)$ certainly satisfies (3), by problem 2 of Exercise 7, and so for each n the function

$$u_n(r, \theta) = \frac{1}{2\pi} \int_0^{2\pi} \tau_n(\varphi) \frac{1 - r^2}{1 + r^2 - 2r\cos(\varphi - \theta)} d\varphi$$

is harmonic in G, continuous on $G \cup T$, and equal to $\tau_n(\theta)$ on T. It follows, because $\lim \tau_n(\theta) = f(\theta)$ uniformly over T (Exercises 4, problem 1) that the sequence $\{u_n(r, \theta)\}$ converges uniformly over $G \cup T$ to a function $u(r, \theta)$ that is continuous on this set and satisfies $u(1, \theta) = f(\theta)$ on T. But clearly, in G,

$$u(r, \theta) = \frac{1}{2\pi} \int_0^{2\pi} f(\varphi) \frac{1 - r^2}{1 + r^2 - 2r\cos(\varphi - \theta)} d\varphi \tag{4}$$

because $\{\tau_n\}$ converges to $f(\theta)$ uniformly over $[0, 2\pi]$. Now u is continuous on $G \cup T$, $u = f$ on T, and u is given by (4) in G, hence

$$\lim_{r \to 1^-} \frac{1}{2\pi} \int_0^{2\pi} f(\varphi) \frac{1 - r^2}{1 + r^2 - 2r\cos(\varphi - \theta)} d\varphi = f(\theta) \ for\ each\ \theta \in T.$$

Exercises 8

1. Equation (4) is called Poisson's integral. The family of functions $P_r(\theta) = \frac{1-r^2}{1+r^2-2r\cos\theta}$, $0 \leq r \leq 1$ is called Poisson's kernel. One should compare (4) with equation (6) of Section 6. See also Lemma 1 of that section. Show that

    (a) $P_r(\theta) \geq 0$ and P_r is continuous on T for all $r \in [0,1]$.

    (b) $\frac{1}{2\pi} \int_0^{2\pi} P_r(\theta) d\theta = 1$ for $0 \leq r \leq 1$.

    (c) If $0 < \delta < \pi$, then $\lim_{r \to 1} \sup_{|\theta| \geq \delta} |P_r(\theta)| = 0$

    Hint: If $\delta \leq |\theta| \leq \pi$, then $P_r(\theta) \leq \frac{1-r^2}{1+r^2-2r\cos\delta}$.

2. Verify that $P_r(\theta)$ is harmonic in G.

3. The formula (4) enables us to write down the solution to the DP for a circle of radius R centered at the origin. If f is continuous on the circumference of this circle, then setting $r = \frac{r}{R}$ and $s = R\varphi$ gives

$$u(r,\theta) = \frac{1}{2\pi R} \int_0^{2\pi R} f(s) \frac{R^2 - r^2}{R^2 + r^2 - 2Rr\cos(\varphi - \theta)} ds$$

Notes

1. The work of D'Alembert, Euler, and Bernoulli on the vibrating string problem is discussed in *Trigonometric Series—A Survey*, by R. L. Jeffery (University of Toronto Press, Toronto, 1965).

2. A discussion of Riemann's function, and the mysteries surrounding its origin, can be found in P. L. Butzer and E. L. Stark, "Riemann's example of a continuous nondifferentiable function in the light of two letters (1865) of Christoffel to Prym," Bull. Soc. Math.Belg. 38 (1986), pp. 45–73.

    J. Gerver published two papers about this function:

    "The differentiability of the Riemann function at certain rational multiples of π," Amer. J. Math. 92 (1970), pp. 33–55.

    "More on the differentiability of the Riemann function," Amer. J. Math. 93 (1970), pp. 33–41.

3. The Fermi paradox is the subject of a book, *Where Is Everybody?*, by Stephen Webb (Copernicus Books in association with Praxis Publishing, Ltd., 2002). See especially pages 17–18 and 28–29.

Chapter Two
Extensions of the Classical Theory

Fourier series are used to solve many problems arising in engineering and in the physical sciences. The functions encountered here are generally real-valued and defined only on some interval of finite length. They are not periodic and often have several points of discontinuity. To discuss the mathematics behind these applications, we must reformulate the definitions given in the last chapter to, somehow, include more general functions.

It was the French physicist Fourier who stated, in 1807, that an arbitrary function could be represented by a trigonometric series provided the coefficients in that series were computed from the Euler formulas. He offered no proof of this, but he used these series extensively to solve physical problems and his work brought harmonic analysis to the attention of his colleagues. This is why these series now bear his name.

1. Functions on $(-\pi, \pi)$

The Euler formulas imply that any function we use them on must be integrable. We are supposing that the reader is familiar with the Riemann integral and most, perhaps all, functions arising in applications are integrable in this sense. So throughout this chapter "function" means "Riemann integrable function."

Definition 1. Let the function f be defined on $(-\pi, \pi)$. Then the Fourier coefficients of f are

$$a_n = \frac{1}{\pi} \int_{-\pi}^{\pi} f(x) \cos nx \, dx \, , \, n = 0, 1, 2, \ldots \tag{1}$$

$$b_n = \frac{1}{\pi} \int_{-\pi}^{\pi} f(x) \sin nx \, dx \, , \, n = 1, 2, 3, \ldots \tag{2}$$

The (real) Fourier series of f is

$$\frac{1}{2} a_0 + \sum_{n=1}^{\infty} (a_n \cos nx + b_n \sin nx) \tag{3}$$

It is customary to write

$$f(x) \sim \frac{1}{2} a_0 + \sum_{n=1}^{\infty} (a_n \cos nx + b_n \sin nx)$$

to indicate that the series is the Fourier series of the given function, i.e., that the a_n's and b_n's were computed from $f(x)$ using the Euler formulas.

We can define a new function, $\tilde{f}(x)$, by setting

$$\tilde{f}(x) = \frac{1}{2} a_0 + \sum_{n=1}^{\infty} (a_n \cos nx + b_n \sin nx)$$

at every x at which the series is convergent. The hope, of course, is that $\tilde{f}(x)$ equals $f(x)$ for each $x \in (-\pi, \pi)$. Now if the series does converge, pointwise, on $(-\pi, \pi)$, then because sine and cosine are 2π-periodic, the series converges everywhere except, perhaps, at odd multiples of π. Thus $\tilde{f}$ is defined at all points of $\mathbb{R}$ except, perhaps, at the points mentioned. Moreover, $\tilde{f}$ is 2π-periodic, meaning $\tilde{f}(x + 2\pi) = \tilde{f}(x)$ all $x \notin \{n\pi \mid n \in Z, n \text{ odd}\}$. Thus, unless f is 2π periodic, the best we can hope for is that $f(x) = \tilde{f}(x)$ in $(-\pi, \pi)$ or $[-\pi, \pi]$, and the following examples show this clearly.

Example 1

Consider the function $f(x) = x^2$ on $(-\pi, \pi)$. We have seen that

$$f(x) \sim \frac{\pi^2}{3} + 4 \sum_{n=1}^{\infty} \frac{(-1)^n}{n^2} \cos nx$$

and the series converges, uniformly in fact, for all x (Chapter One, Exercises 5, problem 1). Thus

$$\tilde{f}(x) = \frac{\pi^2}{3} + 4\sum \frac{(-1)^n}{n^2} \cos nx$$

is a 2π-periodic function on $\mathbb{R}$, a function in $C(T)$.

It is tempting to say that the series converges to x^2, but this is true only on $[-\pi, \pi]$ (as we saw in Chapter One, Section 7, Remark 1, this particular series converges to x^2 at both π and $-\pi$). If we set $x = 2\pi$, the series converges to $\tilde{f}(2\pi) = \tilde{f}(0+2\pi) = \tilde{f}(0) = 0$ and not to $f(2\pi) = 4\pi^2$. The graph of $f(x) = x^2$ is well-known; the graph of $\tilde{f}$ is shown below:

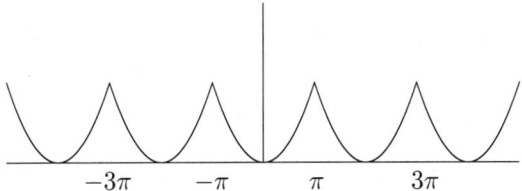

■

Example 2

Let $f(x) = x$ for all $x \in (-\pi, \pi)$. Then

$$f(x) \sim 2 \sum_{n=1}^{\infty} \frac{(-1)^{n+1}}{n} \sin nx$$

as is easily seen. We shall see that this series actually converges at each point of $(-\pi, \pi)$ and that its limit, $\tilde{f}(x)$, is equal to $f(x)$ on this interval. But it is clear that the series converges at every odd multiple of π as well and that its limit is zero. Thus $\tilde{f}(n\pi) = 0$ for all odd n (also for even n in this case). The graph of $\tilde{f}$ is shown below:

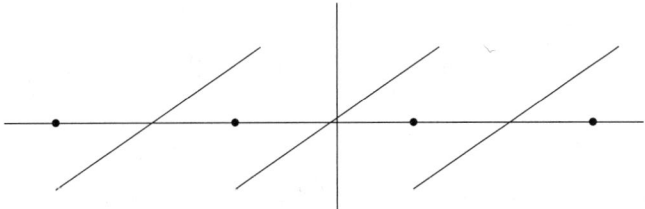

The dots are on the x-axis at $\pm\pi$, $\pm 3\pi$, $\pm 5\pi$, etc.

■

Recall that a function f defined on $(-a, a)$ is said to be an even function if $f(-x) = f(x)$ for all x in this interval. It is easy to see that, in this case,

$$\int_{-a}^{a} f(x)dx = 2 \int_{0}^{a} f(x)dx$$

A function $g(x)$ on $(-a, a)$ is said to be an odd function if $g(-x) = -g(x)$ for all x in this interval. Again one can easily see that, in this case,

$$\int_{-a}^{a} g(x)dx = 0$$

Now when f is an even function on $(-\pi, \pi)$, then $f(x) \cos nx$ is also an even function for all $n = 0, 1, 2, \ldots$. Hence,

$$a_n = \frac{1}{\pi} \int_{-\pi}^{\pi} f(x) \cos nx \, dx = \frac{2}{\pi} \int_{0}^{\pi} f(x) \cos nx \, dx, \quad \text{for each } n$$

Also, because $f(x) \sin nx$ is an odd function for all $n = 1, 2, \ldots$,

$$b_n = \frac{1}{\pi} \int_{-\pi}^{\pi} f(x) \sin nx \, dx = 0 \quad \text{for each } n$$

Similarly, when $g(x)$ is an odd function on $(-\pi, \pi)$ we have

$$a_n = \frac{1}{\pi} \int_{-\pi}^{\pi} g(x) \cos nx \, dx = 0, \quad n = 0, 1, 2, \cdots$$

and

$$b_n = \frac{1}{\pi} \int_{-\pi}^{\pi} g(x) \sin nx \, dx = \frac{2}{\pi} \int_{0}^{\pi} g(x) \sin nx \, dx, \quad n = 1, 2, \cdots$$

Examples 1 and 2 illustrate these results. The function $f(x) = x^2$ is even on $(-\pi, \pi)$. We have seen that its Fourier series contains only even functions, every $b_n = 0$, and that

$$\frac{1}{2}a_0 = \frac{\pi^2}{3}, \quad \text{so } a_0 = 2\left(\frac{\pi^2}{3}\right) \quad \text{and} \quad a_n = 2\left[\frac{2(-1)^n}{n^2}\right], \quad n = 1, 2, \ldots$$

The function $f(x) = x$ is an odd function on $(-\pi, \pi)$. Its Fourier series contains only odd functions, each $a_n = 0$ and $b_n = 2\frac{(-1)^{n+1}}{n}$, $n = 1, 2, \cdots$.

Exercises 1

1. (a) Show that the Fourier series of $f(x) = x$ on $(-\pi, \pi)$ is

$$2 \sum_{n=1}^{\infty} \frac{(-1)^{n+1}}{n} \sin nx$$

(b) Show that the Fourier series of $f(x) = e^x$ on $(-\pi, \pi)$ is

$$\frac{2\sinh\pi}{\pi}\left[\frac{1}{2} + \sum_{n=1}^{\infty}\frac{(-1)^n}{1+n^2}(\cos nx - n\sin nx)\right]$$

(c) Let $f(x) = 1$ when $-\pi < x < 0$, and let $f(x) = 2$ when $0 < x < \pi$. Show that $f \sim \dfrac{3}{2} + \dfrac{1}{\pi}\sum_{n=1}^{\infty}\dfrac{1-(-1)^n}{n}\sin nx$.

(d) Let $g(x) = 0$ when $-\pi < x < 0$, and let $g(x) = \sin x$ when $0 < x < \pi$. Show that $g \sim \dfrac{1}{\pi} + \dfrac{1}{2}\sin x - \dfrac{2}{\pi}\sum_{n=1}^{\infty}\dfrac{\cos 2nx}{4n^2-1}$.

2. (a) Let f be an even function on $(-a, a)$ and let g be an odd function on this interval. Show that

$$\int_{-a}^{a} f(x)dx = 2\int_{0}^{a} f(x)dx \quad \text{and} \quad \int_{-a}^{a} g(x)dx = 0$$

(b) Show that $f \cdot g(x) = f(x)g(x)$ is an odd function.

3. Give f defined on $\mathbb{R}$ set $f_o(x) = \dfrac{f(x) - f(-x)}{2}$ and set $f_e(x) = \dfrac{f(x) + f(-x)}{2}$. Observe that $f(x) = f_e(x) + f_o(x)$.

(a) Show that f_o is an odd function and f_e is an even function.
(b) Compute f_e and f_o for $f(x) = e^x$.

2. Functions on Other Intervals

Suppose now that our function is given only on $(-a, a)$. By analogy with what we did in the last section, we would expect that the Fourier series of the given function should converge to a $2a$-periodic function. Now the period of $\sin bx$ is $\frac{2\pi}{b}$. This is $2a$ when $b = \pi/a$. So we would expect our series to be of the form

$$\frac{1}{2}a_0 + \sum_{n=1}^{\infty}\left[a_n \cos\left(\frac{n\pi x}{a}\right) + b_n \sin\left(\frac{n\pi x}{a}\right)\right] \tag{1}$$

Given a function f on $(-a, a)$, we define the Fourier coefficients of f to be the numbers

$$a_n = \frac{1}{a}\int_{-a}^{a} f(x)\cos\left(\frac{n\pi x}{a}\right)dx, \quad n = 0, 1, 2, \cdots$$

and

$$b_n = \frac{1}{a}\int_{-a}^{a} f(x)\sin\left(\frac{n\pi x}{a}\right)dx, \quad n = 1, 2, 3, \cdots$$

Having found these numbers we define the Fourier series of f to be (1).

As an example, consider $f(x) = x^2$ on $(-a, a)$. We find that

$$f(x) \sim \frac{a^2}{3} + \frac{4a^2}{\pi^2} \sum_{n=1}^{\infty} \frac{(-1)^n}{n^2} \cos\left(\frac{n\pi x}{a}\right)$$

When $a = \pi$ we obtain the series found in the last section (Example 1). It is clear that this series converges uniformly over $\mathbb{R}$ to a function $\tilde{f}$ that is continuous and $2a$-periodic (i.e., $2a$ is a period of this function).

Recall that Fejér's theorem tells us that any function in $C(T)$ is the uniform limit of the Cesàro means of its Fourier series. Given a continuous $2a$-periodic function f we may regard f as a function on the circle with radius a/π centered at the origin. A change of variable in Fejér's theorem tells us that any such f is the uniform limit of the Cesàro means of its Fourier series (i.e., the series (1)).

Very often the functions arising in applications are defined only on an interval of the form $(0, a)$. Given $f(x)$ on this interval there are two, among many, natural ways to extend it to $(-a, a)$. We may define

$$f_e(x) = \begin{cases} f(x) & \text{for } 0 < x < a \\ f(-x) & \text{for } -a < x < 0 \end{cases}$$

This is called the even extension of f. The Fourier series of this function is called the Fourier cosine series for f. We have

$$f_e(x) \sim \frac{1}{2} a_0 + \sum_{n=1}^{\infty} a_n \cos\left(\frac{n\pi x}{a}\right)$$

where

$$a_n = \frac{2}{a} \int_0^a f(x) \cos\left(\frac{n\pi x}{a}\right) dx, \quad n = 0, 1, 2, \cdots$$

If this series converges, its limit will be a $2a$-periodic function, the periodic extension of $f_e(x)$.

We may also define

$$f_o(x) = \begin{cases} f(x) & \text{for } 0 < x < a \\ -f(-x) & \text{for } -a < x < 0 \end{cases}$$

This is called the odd extension of f and its Fourier series

$$f_o(x) \sim \sum_{n=1}^{\infty} b_n \sin\left(\frac{n\pi x}{a}\right)$$

where
$$b_n = \frac{2}{a}\int_0^a f_\theta(x) \sin\left(\frac{n\pi x}{a}\right) dx, \quad n = 1, 2, 3, \cdots$$
is called the Fourier sine series for f. Again, if the series converges, its limit will be the $2a$-periodic extension of $f_\theta(x)$.

These series are called the half-range expansions of $f(x)$. Let us give some examples.

Example 1

Let $f(x) = x$ on $(0, \pi)$. Then $f_\theta(x) = x$ on $(-\pi, \pi)$; technically, f_θ is not defined at 0, but this does not effect its Fourier series. That series is the one found earlier, specifically,
$$f_\theta(x) \sim 2\sum_{n=1}^{\infty} \frac{(-1)^{n+1}}{n} \sin nx$$

The even extension of $f(x)$, $f_e(x)$, coincides with $|x|$ (except, again, at $x = 0$). Here we have
$$f_e(x) \sim \frac{\pi}{2} - \frac{4}{\pi}\sum_{n=1}^{\infty} \frac{\cos(2n-1)x}{(2n-1)^2}$$
∎

Example 2

Suppose we modify Example 1 by taking $g(x) = x$ on $(0, 1)$. Then $g_\theta(x) = x$ on $(-1, 1)$ and
$$b_n = 2\int_0^1 x \sin n\pi x \, dx = -\frac{2}{n\pi}\cos n\pi = \frac{2(-1)^{n+1}}{n\pi}$$
Thus
$$g_\theta(x) \sim \frac{2}{\pi}\sum_{n=1}^{\infty} \frac{(-1)^{n+1}}{n} \sin n\pi x$$
∎

The even extension, $g_e(x)$, coincides with $|x|$ on $(-1, 1)$, giving us
$$a_0 = 2\int_0^1 x \, dx = 1$$
$$a_n = 2\int_0^1 x \cos n\pi x \, dx = \frac{-2}{n^2\pi^2}(1 - \cos n\pi) = \frac{-2(1 - (-1)^n)}{n^2\pi^2}$$

and
$$g_e(x) \sim \frac{1}{2} - \sum_{n=1}^{\infty} \frac{2[1-(-1)^n]}{n^2\pi^2} \cos n\pi x$$

Exercises 2

1. (a) Let $f(x) = -1$ when $-a < x < 0$, $f(x) = 1$ when $0 < x < a$. Show that
$$f \sim \frac{4}{\pi} \sum_{n=1}^{\infty} \left(\frac{1}{2n-1}\right) \sin\left[\frac{(2n-1)\pi x}{a}\right]$$

   (b) Let $f(x) = x + x^2$ for $-1 < x < 1$. Show that
$$f \sim \frac{1}{3} + \frac{2}{\pi} \sum_{n=1}^{\infty} (-1)^n \left[\frac{2}{\pi n^2} \cos n\pi x - \frac{1}{n} \sin n\pi x\right]$$

   (c) Let $f(x) = 0$ when $-2 < x < 1$, $f(x) = 1$ when $1 < x < 2$. Show that
$$f \sim \frac{1}{4} - \frac{1}{\pi} \sum_{n=1}^{\infty} \frac{1}{n} \left[\sin \frac{n\pi}{2} \cos \frac{n\pi x}{2} + \left(\cos n\pi - \cos \frac{n\pi}{2}\right) \sin \frac{n\pi x}{2}\right]$$

2. Here we compute some half-range expansions:

   (a) If $f(x) = \sin x$ when $0 < x < \pi$, then $f_\theta(x) = \sin x$ and
$$f_e(x) \sim \frac{2}{\pi} - \frac{4}{\pi} \sum_{n=1}^{\infty} \frac{\cos 2nx}{4n^2 - 1}$$

   (b) If $f(x) = \cos x$ when $0 < x < \pi$, then
$$f_\theta(x) \sim \frac{8}{\pi} \sum_{n=1}^{\infty} \frac{n \sin 2nx}{4n^2 - 1}, \quad f_e(x) = \cos x$$

   (c) If $f(x) = e^x$ on $(0, \pi)$, then $f_\theta(x) \sim \frac{2}{\pi} \sum_{n=1}^{\infty} [1 - (-1)^n e^\pi] \frac{n \sin nx}{n^2 + 1}$
   and $f_e(x) \sim \frac{e^\pi - 1}{\pi} - \frac{2}{\pi} \sum_{n=1}^{\infty} [1 - (-1)^n e^\pi] \frac{\cos nx}{n^2 + 1}.$

   (d) If $f(x) = \pi - x$ on $(0, \pi)$, then
$$f_\theta(x) \sim 2 \sum_{n=1}^{\infty} \frac{\sin nx}{n}, \quad f_e(x) \sim \frac{\pi}{2} + \frac{4}{\pi} \sum_{n=1}^{\infty} \frac{\cos(2n-1)x}{(2n-1)^2}$$

3. Functions With Special Properties

The functions that arise in the applications of Fourier analysis to engineering are, as illustrated by the examples given above, usually continuous except at a finite number of points. Moreover, at these exceptional points the functions often have one-sided derivatives (defined below). Here we shall investigate the relevant properties of functions of this type.

Let a, b be real numbers with $a < b$. A function f that is defined and continuous on (a, b) and for which the limits, $\lim_{h \to 0^+} f(a+h)$ and $\lim_{h \to 0^+} f(b-h)$, both exist (i.e., they are real numbers) can obviously be extended to a function $\bar{f}$ that is defined and continuous on $[a, b]$; just take $\bar{f}(x) = f(x)$ in (a, b) and set $\bar{f}(a)$ equal to the first of these limits and $\bar{f}(b)$ equal to the second one. Because $\bar{f}$ is continuous on $[a, b]$, it can be approximated by a polynomial on that interval (Chapter One, Section 7, Corollary 3). So it follows that f is bounded on (a, b) and, given $\varepsilon > 0$, there is a polynomial $P_\varepsilon(x)$ such that

$$|f(x) - P_\varepsilon(x)| < \varepsilon$$

for all $x \in (a, b)$.

We can characterize these functions another way. Recall (Chapter Zero, Section 5, Definition 3):

Definition 1. The function $g(x)$ defined on (a, b) is uniformly continuous on that interval if, given $\varepsilon > 0$, there is a $\delta > 0$ such that $|g(x) - g(y)| < \varepsilon$ whenever $x, y \in (a, b)$ and $|x - y| < \delta$.

It is easy to see that if g is uniformly continuous on (a, b), $\lim_{h \to 0^+} g(a+h)$ and $\lim_{h \to 0^+} g(b-h)$ both exist, and conversely, if f is continuous on (a, b) and these one-sided limits exist for f, then f is uniformly continuous on (a, b) (see Exercises 3, problem 4).

Definition 2. A function f defined on (a, b) is said to be piecewise continuous, or sectionally continuous, on this interval if there is a finite (perhaps empty) set $\{x_j\}_{j=1}^{n} \subseteq (a, b)$ such that f is discontinuous at each x_j and f is uniformly continuous on each of the intervals (x_j, x_{j+1}), $j = 0, \ldots, n$ where $x_0 = a$, $x_{n+1} = b$. We shall call $\{x_j\}_{j=0}^{n+1}$ the subdivision of f.

Given any $y \in (a, b)$ we set

$$f(y+0) = \lim_{h \to 0^+} f(y+h)$$

and

$$f(y-0) = \lim_{h \to 0^+} f(y-h)$$

When f is piecewise continuous on (a, b) these limits always exist. For any such function the numbers

$$f(a + 0) = \lim_{h \to 0+} f(a + h)$$

and

$$f(b - 0) = \lim_{h \to 0+} f(b - h)$$

also exist.

Theorem 1. If f is piecewise continuous on (a, b), then $\lim_{\lambda \to \infty} \int_a^b f(x) \sin \lambda x \, dx = 0$ and $\lim_{\lambda \to \infty} \int_a^b f(x) \cos \lambda x \, dx = 0$.

Proof. It is clearly sufficient to prove this result for a function f that is uniformly continuous on (a, b). Now given $\varepsilon > 0$ we must show that

$$\left| \int_a^b f(x) \cos \lambda x \, dx \right| < \varepsilon$$

for all $\lambda \geq \Lambda$; i.e., given $\varepsilon > 0$ we must show that we can find Λ such that this inequality is true. Now we may choose a polynomial $P_\varepsilon(x)$ such that

$$\int_a^b |f(x) - P_\varepsilon(x)| dx < \frac{\varepsilon}{2}$$

for all we need do is choose $P_\varepsilon(x)$ so that

$$|f(x) - P_\varepsilon(x)| < \frac{\varepsilon}{2(b-a)}$$

for all $x \in (a, b)$. Clearly then

$$\left| \int_a^b f(x) \cos \lambda x \, dx \right| \leq \left| \int_a^b f(x) \cos \lambda x \, dx - \int_a^b P_\varepsilon(x) \cos \lambda x \, dx \right| + \left| \int_a^b P_\varepsilon(x) \cos \lambda x \, dx \right|$$

$$\leq \int_a^b |f(x) - P_\varepsilon(x)| |\cos \lambda x| dx + \left| \int_a^b P_\varepsilon(x) \cos \lambda x \, dx \right|$$

$$\leq \frac{\varepsilon}{2} + \left| \int_a^b P_\varepsilon(x) \cos \lambda x \, dx \right|$$

In this last integral we may integrate by parts. We find

$$\int P_\varepsilon(x) \cos \lambda x \, dx = \frac{P_\varepsilon(x) \sin \lambda x}{\lambda} - \frac{1}{\lambda} \int P'_\varepsilon(x) \sin \lambda x \, dx$$

and so

$$\left| \int_a^b P_\varepsilon(x) \cos \lambda x \, dx \right| \leq \left| \frac{P_\varepsilon(b) \sin \lambda b - P_\varepsilon(a) \sin \lambda a}{\lambda} \right| + \frac{1}{\lambda} \left| \int_a^b P'_\varepsilon(x) \sin \lambda x \, dx \right|$$

We can choose Λ_1, so that the first term is less than $\varepsilon/4$ when $\lambda \geq \Lambda_1$. The term containing the integral can be estimated as follows:

$$\frac{1}{\lambda} \left| \int_a^b P'_\varepsilon(x) \sin \lambda x \, dx \right| \leq \sup_{a < x < b} |P'_\varepsilon(x)| \frac{(b-a)}{\lambda}$$

where we recall that $P_\varepsilon(x)$ is a polynomial; hence, $P'_\varepsilon(x)$ is uniformly continuous on (a, b) and, in particular, bounded on that interval. Thus we can find Λ_2 such that this integral is less than $\varepsilon/4$ whenever $\lambda \geq \Lambda_2$. Clearly then

$$\left| \int_a^b f(x) \cos \lambda x \, dx \right| < \varepsilon$$

whenever $\lambda \geq \max\{\Lambda_1, \Lambda_2\}$, and the proof is complete.

A similar argument shows that

$$\lim_{\lambda \to \infty} \int_a^b f(x) \sin \lambda x \, dx = 0$$

Corollary 1. The Fourier coefficients of a piecewise continuous function tend to zero as $n \to \infty$. This generalizes Riemann's theorem (Chapter One, Section 5, Theorem 1).

We recall that if $x_0 \in (a,b)$, we say that f is differentiable at this point provided
$$\lim_{h \to 0} \frac{f(x_0+h) - f(x_0)}{h}$$
exists. We denote this limit by $f'(x_0)$. Now
$$f(x_0+h) = \left[\frac{f(x_0+h) - f(x_0)}{h}\right] h + f(x_0)$$
and so, when $f'(x_0)$ exists, $\lim_{h \to 0} f(x_0+h) = f(x_0)$; i.e., differentiability of f at x_0 implies the continuity of f at that point. We shall now define the one-sided derivatives of f obtaining a useful generalization of the derivative that can exist even at discontinuities of f. We shall use the notation introduced just after Definition 2.

Definition 3. Let f be piecewise continuous on (a,b), and let $x_0 \in (a,b)$. Then the right-hand derivative of f at x_0, $f'_+(x_0)$, is defined to be
$$\lim_{h \to 0+} \frac{f(x_0+h) - f(x_0+0)}{h}$$
provided that this limit exists. We define the left-hand derivative of f at x_0, $f'_-(x_0)$, to be
$$\lim_{h \to 0+} \frac{f(x_0-h) - f(x_0-0)}{h}$$
provided that this limit exists.

It is clear that when f has a derivative at x_0, then both its right- and left-hand derivatives exist and both are equal to $f'(x_0)$. The function $f(x) = 2x$, $x \leq 1$, $f(x) = 1-x$, $1 < x$, however, is discontinuous, hence nondifferentiable, at $x = 1$, but $f'_+(1)$ and $f'_-(1)$ both exist. The first of these is -1; the second is 2.

Lemma 1. If the two functions f, g both have a right-hand derivative (or a left-hand derivative) at a point x_0, then also so does $f(x) \cdot g(x)$.

Proof. For any $h > 0$ we may write
$$\frac{f(x_0+h)g(x_0+h) - f(x_0+0)g(x_0+0)}{h}$$
$$= f(x_0+h)\left[\frac{g(x_0+h) - g(x_0+0)}{h}\right] + g(x_0+0)\left[\frac{f(x_0+h) - f(x_0+0)}{h}\right]$$
and the result follows from this.

Not surprisingly, we can define the right-hand derivative at a and the left-hand derivative at b for a piecewise continuous function f on (a, b). They are

$$f'_+(a) = \lim_{h \to 0^+} \frac{f(a+h) - f(a+0)}{h}$$

and

$$f'_-(b) = \lim_{h \to 0^+} \frac{f(b-h) - f(b-0)}{h}$$

We can give conditions on f that imply the existence of these derivatives.

Lemma 2. Let f be uniformly continuous on (a, b). Suppose that $f'(x)$ exists at every point of this interval and that f' is uniformly continuous here. Then the right- and left-hand derivatives of f at a and b, respectively, exist. Furthermore, $f'_+(a) = f'(a+0)$ and $f'_-(b) = f'(b-0)$.

Proof. We may define a continuous function $\bar{f}$ on $[a, b]$ by setting $\bar{f}(x) = f(x)$ when $x \in (a, b)$, $\bar{f}(a) = f(a+0)$, $\bar{f}(b) = f(b-0)$. Clearly, $\bar{f}$ has a derivative at each point of (a, b); in fact, $\bar{f}'(x) = f'(x)$ and every point of (a, b). By the mean value theorem (Chapter Zero, Section 5, Theorem 5) we may write

$$\frac{\bar{f}(a+h) - \bar{f}(a)}{h} = \bar{f}'(a + \theta h)$$

where $h > 0$, and θ, $0 < \theta < 1$, depends on h. But this is just $\frac{f(a+h) - f(a+0)}{h} = f'(a + \theta h)$ and, letting $h \to 0^+$ proves the result at $x = a$. The proof for the point $x = b$ is similar.

Corollary 1. If f is piecewise continuous on (a, b) with subdivision $\{x_j\}_{j=0}^{n+1}$ (see Definition 2), and if f' is piecewise continuous on (a, b) with the same subdivision, then the one-sided derivatives of f exist at every point of $[a, b]$.

Exercises 3

*1. Show that $f(x) = \frac{x}{2\sin(\frac{1}{2}x)}$ has both a left- and a right-hand derivative at $x = 0$. What is $\lim_{x \to 0} f(x)$?

2. Let $f(x) = x$ on $(-\pi, \pi)$. Let $f(x + 2\pi) = f(x)$. Compute:

   (a) $f(\pi - 0)$;   (b) $f(\pi + 0)$;   (c) $f'(\pi - 0)$;   (d) $f'(\pi + 0)$.

3. Let $f(x) = x$ if $0 \leq x < \frac{\pi}{2}$, $f(x) = \pi - x$ if $\frac{\pi}{2} \leq x \leq \pi$. Compute:

   (a) $f(\frac{\pi}{2} - 0)$;   (b) $f(\frac{\pi}{2} + 0)$;   (c) $f'(\frac{\pi}{2} - 0)$;   (d) $f'(\frac{\pi}{2} + 0)$.

*4. Some authors define f to be piecewise continuous on (a,b) if there is a finite (perhaps empty) set $\{x_j\}_{j=1}^n$ such that f is continuous on (x_j, x_{j+1}), $j = 0, 1, \ldots, n$, where $a = x_0$ and $b = x_{n+1}$, and $\lim_{x \to x_j^+} f(x)$, $\lim_{x \to x_j^-} f(x)$ exist for each $j = 1, 2, \ldots n$, $\lim_{x \to a^+} f(x)$ exists and $\lim_{x \to b^-} f(x)$ exists. Show that this is equivalent to our Definition 2 as follows. Clearly, it suffices to prove the result when $\{x_j\} = \phi$.

(a) If f is continuous on (a,b) and $\lim_{x \to a^+} f(x)$, $\lim_{x \to b^-} f(x)$ both exist, define $\bar{f}$ on $[a,b]$ by letting $\bar{f}(x) = f(x)$ when $x \in (a,b)$, $\bar{f}(a) = \lim_{x \to a^+} f(x)$, $\bar{f}(b) = \lim_{x \to b^-} f(x)$. Then $\bar{f}$ is continuous on the closed bounded interval $[a,b]$. Now use Theorem 3 of Chapter Zero, Section 5 to show that f is uniformly continuous on (a,b).

(b) Now suppose that f is uniformly continuous on (a,b). We must show that $\lim_{x \to a^+} f(x)$ and $\lim_{x \to b^-} f(x)$ both exist.

i. If $\{x_n\} \subseteq (a,b)$ is a Cauchy sequence (Chapter Zero, Exercises 5, Definition 5), show that $\{f(x_n)\}$ is a Cauchy sequence.

ii. Choose $\{x_n\} \subseteq (a,b)$ such that $\lim_{n \to \infty} x_n = a$ and let $\alpha = \lim_{n \to \infty} f(x_n)$. Given $\varepsilon > 0$, find $\delta > 0$ such that $|\alpha - f(a+h)| < \varepsilon$ whenever $0 < h < \delta$. Thus $\lim_{x \to a^+} f(x) = \alpha$. The limit of f at b can be treated similarly.

4. Pointwise Convergence of the Fourier Series

In proving Fejér's theorem we made use of an integral representation of the Cesàro means of the Fourier series. This involved a sequence of functions that we called Fejér's kernel. Recall

$$\sigma_m(x;f) = \frac{1}{2\pi} \int_0^{2\pi} K_m(x-y)f(y)dy$$

where

$$K_m(x) = \frac{1}{m}\left[\frac{1-\cos mx}{1-\cos x}\right] = \frac{1}{m}\left(\frac{\sin \frac{1}{2}mx}{\sin \frac{1}{2}x}\right)^2$$

Here we shall derive an integral representation for the partial sums of a Fourier series. This, too, involves a sequence of functions that we shall call Dirichlet's kernel. We find that

$$s_n(x;f) = \frac{1}{\pi}\int_{-\pi}^{\pi} f(t)\, D_n(t-x)dt$$

where
$$D_n(x) = \frac{\sin(n + \frac{1}{2})x}{2\sin\frac{1}{2}x}$$

We begin with a lemma due to Lagrange.

Lemma 1. $2\sum_{k=1}^{n} \cos k\theta = -1 + \dfrac{\sin(n+\frac{1}{2})\theta}{\sin\frac{1}{2}\theta}.$

Proof. $2(\cos\theta + \cos 2\theta + \cdots + \cos n\theta) = \sum_{k=1}^{n} e^{ik\theta} + \sum_{k=1}^{n} e^{-ik\theta}.$ We may sum each of these finite geometric series to get

$$2\sum_{k=1}^{n}\cos k\theta = \frac{e^{i\theta}(1 - e^{in\theta})}{1 - e^{i\theta}} + \frac{e^{-i\theta}(1 - e^{-in\theta})}{1 - e^{-i\theta}}$$

$$= \frac{e^{i\theta}(1 - e^{in\theta})}{1 - e^{i\theta}} + \frac{(1 - e^{-in\theta})}{e^{i\theta}(1 - e^{-i\theta})}$$

$$= \frac{e^{i\theta}(1 - e^{in\theta})}{1 - e^{i\theta}} - \frac{(1 - e^{-in\theta})}{1 - e^{i\theta}} = \frac{e^{i\theta} - e^{i(n+1)\theta} - 1 + e^{-in\theta}}{1 - e^{i\theta}}$$

In this last quotient we multiply numerator and denominator by $e^{-i\theta/2}$ to get

$$2\sum_{k=1}^{n}\cos k\theta = \frac{e^{i\frac{\theta}{2}} - e^{i(n+\frac{1}{2})\theta} - e^{-i\frac{\theta}{2}} + e^{-i(n+\frac{1}{2})\theta}}{e^{-i\frac{\theta}{2}} - e^{i\frac{\theta}{2}}}$$

$$= -\frac{(e^{-i\theta/2} - e^{i\theta/2})}{e^{-i\frac{\theta}{2}} - e^{i\frac{\theta}{2}}} + \frac{e^{-i(n+\frac{1}{2})\theta} - e^{i(n+\frac{1}{2})\theta}}{e^{-i\frac{\theta}{2}} - e^{i\frac{\theta}{2}}}$$

$$= -1 + \frac{\cos(n+\frac{1}{2})\theta - i\sin(n+\frac{1}{2})\theta - \{\cos(n+\frac{1}{2})\theta + i\sin(n+\frac{1}{2})\theta\}}{\cos\frac{\theta}{2} - i\sin\frac{\theta}{2} - (\cos\frac{\theta}{2} + i\sin\frac{\theta}{2})}$$

$$= -1 - \frac{2i\sin(n+\frac{1}{2})\theta}{-2i\sin\frac{\theta}{2}} = -1 + \frac{\sin(n+\frac{1}{2})\theta}{\sin\frac{1}{2}\theta}$$

and the proof is complete.

Now given a function f on $(-\pi, \pi)$ the nth partial sum of the Fourier series of f can be transformed into an integral as follows:

$$s_n(x; f) = \frac{1}{2}a_0 + \sum_{k=1}^{n}(a_k \cos kx + b_k \sin kx)$$

$$= \frac{1}{2\pi}\int_{-\pi}^{\pi} f(t)dt + \frac{1}{\pi}\sum_{k=1}^{n}(\cos kx \int_{-\pi}^{\pi} f(t)\cos kt\, dt$$

$$+ \sin kx \int_{-\pi}^{\pi} f(t)\sin kt\, dt)$$

$$= \frac{1}{\pi}\int_{-\pi}^{\pi} f(t)\left\{\frac{1}{2} + \sum_{k=1}^{n}(\cos kx \cos kt + \sin kx \sin kt)\right\}dt$$

$$= \frac{1}{\pi}\int_{-\pi}^{\pi} f(t)\left\{\frac{1}{2} + \sum_{k=1}^{n}\cos k(t-x)\right\}dt$$

Using Lemma 1 inside the curly brackets we obtain

$$s_n(x; f) = \frac{1}{\pi}\int_{-\pi}^{\pi} f(t)\frac{\sin[(n+\frac{1}{2})(t-x)]}{2\sin[\frac{1}{2}(t-x)]}dt \qquad (1)$$

We shall use (1) to investigate the convergence of the Fourier series of f. There is no loss in generality in assuming that f is 2π-periodic, and we shall assume that from now on (Section 1, just after Definition 1).

In (1) we set $u = t - x$ to obtain

$$s_n(x; f) = \frac{1}{2\pi}\int_{-\pi-x}^{\pi-x} f(u+x)\frac{\sin[(n+\frac{1}{2})u]}{\sin\frac{1}{2}u}du$$

In here we replace $\frac{1}{2\pi}\frac{\sin[(n+\frac{1}{2})u]}{\sin\frac{1}{2}u}$ by $D_n(u)$. The reader can check that the function $D_n(u)$ is 2π-periodic. Thus (Chapter One, Exercises 2, problem 3) our last expression for $s_n(x; f)$ becomes

$$s_n(x; f) = \int_{-\pi}^{\pi} f(u+x)D_n(u)du$$

Lemma 2. For each fixed n the function $D_n(u)$ is even (Section 1, just after Example 2) and, furthermore,

$$\int_{-\pi}^{\pi} D_n(u)du = 2\int_{0}^{\pi} D_n(u)du = 1$$

4. Pointwise Convergence of the Fourier Series

Proof. We shall use Lemma 1. We have

$$\int_{-\pi}^{\pi} D_n(u)du = \frac{1}{\pi}\int_{-\pi}^{\pi} \frac{\sin[(n+\frac{1}{2})u]}{2\sin\frac{1}{2}u} du$$

$$= \frac{1}{\pi}\left[\frac{1}{2}\int_{-\pi}^{\pi} du + \sum_{k=1}^{n}\int_{-\pi}^{\pi} \cos ku\, du\right]$$

$$= \frac{1}{\pi}\left[\frac{1}{2}(2\pi)\right]$$

$$= 1$$

Theorem 1. Let f be a 2π-periodic function that is piecewise continuous on $(-\pi, \pi)$. Then the Fourier series of f converges to

$$\frac{1}{2}[f(x+0) + f(x-0)]$$

at every point x where f has both a left- and a right-hand derivative.

Proof. Let x be a point where f has both a left- and a right-hand derivative. There is no loss in generality in assuming that $x \in (-\pi, \pi)$. We must show that

$$\lim_{n\to\infty} s_n(x; f) = \lim_{n\to\infty} \int_{-\pi}^{\pi} f(u+x)D_n(u)du = \frac{f(x+0) + f(x-0)}{2}$$

and to do this it suffices to show that

$$\lim_{n\to\infty} \int_0^{\pi} f(u+x)D_n(u)du = \frac{f(x+0)}{2},$$

$$\lim_{n\to\infty} \int_{-\pi}^{0} f(u+x)D_n(u)du = \frac{f(x-0)}{2}$$

By lemma 2 the first of these is equivalent to showing that

$$\lim_{n\to\infty} \int_0^{\pi} [f(u+x) - f(x+0)]D_n(u)du = 0$$

This integral can be written as follows:

$$\int_0^{\pi} [f(u+x) - f(x+0)] \frac{\sin[n+\frac{1}{2})u]}{2\pi \sin\frac{1}{2}u} du =$$

$$\frac{1}{2\pi}\int_0^{\pi} \left[\frac{f(u+x) - f(x+0)}{\sin\frac{1}{2}u}\right] \sin[(n+\frac{1}{2})u]du$$

By Theorem 1 of Section 3 this tends to zero as $n \to \infty$, provided the function in the square brackets is piecewise continuous on $(0, \pi)$. The only potential difficulty is that the limit, as $u \to 0^+$, might not exist. This, however, is not really a problem because

$$\lim_{u \to 0^+} \frac{f(u+x) - f(x+0)}{\sin \frac{1}{2}u}$$

$$= \lim_{u \to 0^+} \left[\frac{f(u+x) - f(x+0)}{u} \right] \left(\frac{\frac{u}{2}}{\sin \frac{1}{2}u} \right) 2$$

$$= 2 \lim_{u \to 0^+} \frac{f(u+x) - f(x+0)}{u}$$

and this exists because f has a right-hand derivative at x.

The fact that

$$\lim_{n \to \infty} \int_{-\pi}^{0} [f(u+x) - f(x-0)] D_n(u) du = 0$$

can be proved in a similar way.

Remarks.

(1) At any point x_0 where the function f is continuous and has one-sided derivatives, the series converges to $f(x_0)$. This is true, in particular, at any point where $f(x)$ is differentiable.

(2) An analogous theorem is true for any piecewise continuous, $2a$-periodic function (see Section 2 for the definition of the Fourier series of such a function).

(3) If f is defined only on $(-\pi, \pi)$, then the theorem applies to the periodic extension of f to all of $\mathbb{R}$. Hence, if f is piecewise continuous, its Fourier series converges to $\frac{1}{2}[f(x+0) + f(x-0)]$ at each interior point (i.e., point of $(-\pi, \pi)$) where both one-sided derivatives exist. At both end points, $x = \pi$ and $x = -\pi$; however, the series converges to $\frac{1}{2}[f(\pi - 0) + f(-\pi + 0)]$ provided f has a right-hand derivative at $x = -\pi$ and a left-hand derivative at $x = \pi$. Thus if the series is to converge to $f(-\pi + 0)$ at $x = -\pi$ or to $f(\pi - 0)$ when $x = \pi$, it is necessary that $f(-\pi + 0) = f(\pi - 0)$.

4. Pointwise Convergence of the Fourier Series

Let us return now to equation (1). Recall

$$s_n(x;f) = \frac{1}{\pi} \int_{-\pi}^{\pi} \frac{f(t)\sin[(n+\tfrac{1}{2})(t-x)]}{2\sin[\tfrac{1}{2}(t-x)]} \, dt \qquad (1)$$

$$= \frac{1}{\pi} \int_{-\pi-x}^{\pi-x} \frac{f(t)\sin(n+\tfrac{1}{2})u}{\sin\tfrac{1}{2}u} \, du$$

$$= \frac{1}{2\pi} \int_{-\pi}^{0} \frac{f(x+u)\sin(n+\tfrac{1}{2})u}{\sin\tfrac{1}{2}u} \, du$$

$$+ \frac{1}{2\pi} \int_{0}^{\pi} \frac{f(x+u)\sin(n+\tfrac{1}{2}u)u}{\sin\tfrac{1}{2}u} \, du$$

where we have set $u = t - x$. In the first integral on the right we set $v = -u$ to obtain

$$\frac{1}{2\pi} \int_{\pi}^{0} \frac{f(x-v)\sin(n+\tfrac{1}{2})v}{\sin\tfrac{1}{2}v}(-dv) = \frac{1}{2\pi} \int_{0}^{\pi} \frac{f(x-v)\sin(n+\tfrac{1}{2})v}{\sin\tfrac{1}{2}v} \, dv$$

Hence (1) becomes

$$s_n(x;f) = \frac{1}{2\pi} \int_{0}^{\pi} \frac{f(x+u)\sin(n+\tfrac{1}{2})du}{\sin\tfrac{1}{2}u} \, du$$

$$+ \frac{1}{2\pi} \int_{0}^{\pi} \frac{f(x-v)\sin(n+\tfrac{1}{2})v}{\sin\tfrac{1}{2}v} \, dv$$

$$= \frac{1}{2\pi} \int_{0}^{\pi} \frac{\sin(n+\tfrac{1}{2})u}{\sin\tfrac{1}{2}u} \{f(x+u) + f(x-u)\} du$$

We can put this into a more illuminating form by first noting that when $f(x) = 1$ for all x we have $a_0 = 2$ and $s_n(x;f) = 1$ for all n. Thus

$$1 = \frac{1}{2\pi} \int_{0}^{\pi} \frac{\sin(n+\tfrac{1}{2})u}{\sin\tfrac{1}{2}u} (2du)$$

Multiplying this by s and subtracting from $s_n(x;f)$ gives us

$$s_n(x;f) - s = \frac{1}{2\pi} \int_{0}^{\pi} \frac{\sin(n+\tfrac{1}{2})u}{\sin\tfrac{1}{2}u} \{f(x+u) + f(x-u) - 2s\} du$$

Thus a necessary and sufficient condition that $s_n(x;f)$ converge to s is that this last integral tends to zero as $n \to \infty$. There are tests for convergence that are based on giving conditions on f for this to be true (see Notes).

Exercises 4

1. In each of the following cases discuss the convergence of the real Fourier series of the function given.

    (a) $f(x) = x$, $-\pi < x < \pi$
    (b) $f(x) = e^x$, $-\pi < x < \pi$
    (c) $f(x) = 1$, when $-\pi < x < 0$ and $f(x) = 2$ when $0 < x < \pi$
    (d) $f(x) = 0$, when $-\pi < x < 0$ and $f(x) = \sin x$ when $0 < x < \pi$

2. The nth partial sum of the Fourier series of a function $f(t)$ oscillates near a discontinuity of this function. This is known as the Gibbs phenomenon or the Gibbs effect. We give an illustration of this in the following exercise.

    (a) Let $f(t) = \pi - t$ for $0 \leq t \leq \pi$, $f(t) = -\pi - t$ for $-\pi \leq t < 0$. Show that the Fourier series for f is $2 \sum_{n=1}^{\infty} \dfrac{\sin nt}{n}$.

    (b) Let us set $f_k(t) = 2 \sum_{n=1}^{k} \dfrac{\sin nt}{n} - (\pi - t)$ and investigate the behavior of f_k for large k near the discontinuity $t = 0$ of f. Note first that $f_k'(t) = \dfrac{\sin(k + \frac{1}{2})t}{\sin \frac{1}{2}t}$ by Lemma 1.

    (c) The first critical point (i.e., the first point where f' is zero) of f_k to the right of $t = 0$ is $t_k = \dfrac{\pi}{k+\frac{1}{2}}$. Show that $f_k(t_k) = \displaystyle\int_0^{t_k} \dfrac{\sin(k + \frac{1}{2})t}{\sin \frac{1}{2}t} dt - \pi$.

    (d) In the integral in part (c) set $y = (k + \frac{1}{2})t$ and use the fact that $\displaystyle\lim_{y \to 0} \dfrac{\sin y}{y} = 1$ to show that $\displaystyle\lim_{k \to \infty} f_k(t_k) = 2 \int_0^{\pi} \dfrac{\sin y}{y} dy - \pi$.

    (e) Numerical evaluation of this last quantity shows it to be $\simeq 0.562$. Now as $k \to \infty$, $t_k \to 0$, but, as we have shown, the difference between $f(t)$ and its kth partial sum (i.e., f_k) does not tend uniformly to zero. However large k is, there are points arbitrarily close to zero at which f_k is near 0.562.

Notes

The work of Fourier is discussed in the book by Jeffery listed in the Notes to Chapter One. The tests mentioned at the end of Section 4 can be found in *The Theory of Functions,* by E. C. Titchmarsh (Oxford University Press, 2nd edition, London, 1939), see pp. 406–410. Further discussions of the engineering applications of Fourier series can be found in *Advanced Engineering Mathematics,* by Erwin Kreyszig (John Wiley & Sons, Inc., 8th edition, New York, 1999), see pp. 525–648.

Chapter Three

Fourier Series in Hilbert Space

There is another way to look at Fourier series. Before we can present this more geometric point of view, we must first review some linear algebra. We begin informally and leave a more systematic discussion to the next section.

We recall that the vector space $\mathbb{R}^n$ has an inner (sometimes called "dot" or "scalar") product. For $\vec{x} = (x_1, \ldots, x_n)$, $\vec{y} = (y_1, \ldots, y_n)$ we have

$$<\vec{x}, \vec{y}> = \sum_{j=1}^{n} x_j y_j \tag{1}$$

In the space $\mathbb{C}^n$ we must modify (1) a little bit. For $\vec{z}_1 = (z_1, \ldots, z_n)$ and $\vec{w}_1 = (w_z, \ldots, w_n)$ we have

$$<\vec{z}, \vec{w}> = \sum_{j=1}^{n} z_j \overline{w}_j \tag{2}$$

where the bar denotes complex conjugation (Chapter Zero, Exercises 4, problem 1).

If we set $\vec{e}_1 = (1, 0, \ldots, 0)$, $\vec{e}_2 = (0, 1, 0, \ldots, 0)$, $\ldots \vec{e}_n = (0, 0, \ldots, 1)$, then clearly $<\vec{e}_j, \vec{e}_k> = 0$ when $j \neq k$, and this product is 1 when $j = k$. Also, for any vector $\vec{u}$,

$$\vec{u} = \sum_{j=1}^{n} <\vec{u}, \vec{e}_j> \vec{e}_j \tag{3}$$

With these results in mind, let us turn now to $C(T)$. Equation (2) leads us to define

$$<f,g> = \int_{-\pi}^{\pi} f(t)\overline{g(t)}dt \quad \text{for any } f,\ g \text{ in } C(T)$$

Note that for any fixed $n \in Z$,

$$<e^{int}, e^{int}> = 2\pi$$

and so the set $\left\{\dfrac{e^{int}}{\sqrt{2\pi}} \big| n \in Z\right\}$ satisfies $\left\langle \dfrac{e^{int}}{\sqrt{2\pi}}, \dfrac{e^{imt}}{\sqrt{2\pi}} \right\rangle = 0$ if $m \neq n$ and is 1 if $n = m$.

Equation (3) leads us to ask if, for $f \in C(T)$, we may write:

$$f = \sum_{n=-\infty}^{\infty} \left\langle f, \frac{e^{int}}{\sqrt{2\pi}} \right\rangle \frac{e^{int}}{\sqrt{2\pi}}$$

$$= \sum_{n=-\infty}^{\infty} \frac{1}{2\pi} <f, e^{int}> e^{int}$$

$$= \sum_{n=-\infty}^{\infty} \left(\frac{1}{2\pi} \int_{-\pi}^{\pi} f(t)e^{-int}dt\right) e^{int}$$

So we are asking if f can be represented by its Fourier series, but we have not specified in just what sense the series is to "represent" the function. Some kind of convergence must be involved, but the convergence we have in mind here is neither pointwise nor uniform. It is a notion of convergence induced on $C(T)$ by its inner product. This will be made clear below.

1. Normed Vector Spaces

To proceed with our new way of looking at Fourier series, we must first say something about infinite dimensional vector spaces. The terminology of linear algebra is, of course, needed here. A brief discussion of this terminology is given in Appendix A.

Let us begin by recalling some conventions. If D is a nonempty subset of $\mathbb{K}$ and f,g are maps from D into $\mathbb{K}$, then $f+g$ is the function defined as follows: $(f+g)(t) = f(t) + g(t)$ for all $t \epsilon D$. Also, when $\lambda \epsilon \mathbb{K}$, the function λf is defined by $(\lambda f)(t) = \lambda f(t)$ for all $t \epsilon D$. When we consider a vector space of functions defined on D (i.e., a set of functions that has the properties listed in Definition 1 of Appendix A), then the values of these functions determine our scalars. So if all of our functions are real-valued, our scalars are real numbers, and when all of our functions are complex-valued, our scalars are complex numbers.

Recall that, in very elementary physics, a vector is defined to be a quantity having both length and direction. In mathematics we first define a vector space and then say that the elements of that space are to be called vectors. We know nothing about them except that they obey certain axioms. This is the basis for the often repeated remark that mathematics is the subject in which we don't know what we are talking about. Many of the vector spaces that arise in analysis, however, have a function defined on them that enables us to assign a "length" to each element of the space. Such a function is called a "norm." The formal definition is this:

Definition 1. Let E be a vector space over $\mathbb{K}$. A nonnegative real-valued function p on E is said to be a norm on E if

(a) $p(\vec{x} + \vec{y}) \leq p(\vec{x}) + p(\vec{y})$ for all $\vec{x}, \vec{y}$ in E;

(b) $p(\lambda \vec{x}) = |\lambda| p(\vec{x})$ for all λ in $\mathbb{K}$ and all $\vec{x} \epsilon E$;

(c) $p(\vec{x}) = 0$ if, and only if, $\vec{x}$ is the zero vector.

When p is a norm on E, it is customary to denote, for each $\vec{x} \epsilon E$, the number $p(\vec{x})$ by $\|\vec{x}\|$. A vector space on which a norm is defined is called a normed space. If we want to emphasize the norm, say $\|\cdot\|$, on E we speak of the normed space $(E, \|\cdot\|)$.

Let us look at a few examples.

(i) For any fixed $n \epsilon \mathbb{N}$ the vector space $\mathbb{R}^n$ consists of all ordered n-tiples of real numbers. So if $\vec{x} \epsilon \mathbb{R}^n$, then $\vec{x} = (x_1, ..., x_n)$. We define $\|\vec{x}\|$ to be

$$\sqrt{\sum_{j=1}^{n} x_j^2}$$

Because $\lambda \vec{x} = (\lambda x_1, ..., \lambda x_n)$ we see that

$$\|\lambda \vec{x}\|_2 = \left[\sum_{j=1}^{n}(\lambda^2 x_j^2)\right]^{\frac{1}{2}} = \left[\lambda^2 \sum_{j=1}^{n} x_j^2\right]^{\frac{1}{2}} = |\lambda| \left[\sum_{j=1}^{n} x_j^2\right]^{\frac{1}{2}} = |\lambda| \|\vec{x}\|_2$$

Also, $\|\vec{x}\|_2 = 0$ if, and only if,

$$\sum_{j=1}^{n} x_j^2 = 0$$

Clearly, this will happen when, and only when, $x_j = 0$ for each j. So the function $\|\cdot\|_2$ has properties (b) and (c) of Definition 1. Property (a), the so-called triangle inequality, is harder to prove. We present the proof after we discuss inner products.

(ii) We can obtain an important generalization of $\mathbb{R}^n$ by considering all "square summable" sequences of real numbers, i.e., all functions $x : \mathbb{N} \to \mathbb{R}$ such that

$$\sum_{n=1}^{\infty} (x(n))^2 < \infty$$

Writing $x(n) = x_n$ for each n, we are considering all sequences $\vec{x} = (x_1, x_2, ...)$ such that

$$\sum_{k=1}^{\infty} x_k^2 < \infty$$

The set of all such sequences is denoted by $\ell^2(\mathbb{N})$ (read "ℓ two of $\mathbb{N}$"). We shall see, once we discuss inner products, that $\ell^2(\mathbb{N})$ is a vector space and that the function

$$\|\vec{x}\|_2 = \left[\sum_{n=1}^{\infty} x_n^2\right]^{\frac{1}{2}}$$

is a norm on this space.

(iii) There are natural analogues of (i) and (ii) for the vector space of ordered n-tuples of complex numbers, $\mathbb{C}^n$, and the vector space of square summable sequences of complex numbers. The latter is denoted by $\ell_c^2(\mathbb{N})$ or, when there is no possibility of confusion, by $\ell^2(\mathbb{N})$. In the first of these, however, we must define $\|\vec{z}\|_2 =$ to be

$$\left[\sum_{j=1}^{n} |z_j|^2\right]^{\frac{1}{2}}$$

and in the second

$$\|\vec{z}\|_2 = \left[\sum_{n=1}^{\infty} |z_n|^2\right]^{\frac{1}{2}} \quad \text{(see Exercises 1, problem 2).}$$

(iv) Of special importance for us is the vector space $\ell^1(\mathbb{Z})$ (read "ℓ one of $\mathbb{Z}$"). It consists of all functions $a : \mathbb{Z} \to \mathbb{C}$ such that

$$\sum_{n=-\infty}^{\infty} |a(n)| < \infty$$

Observe that when a, b are in $\ell^1(\mathbb{Z})$, then $a + b$ is also in this set; for

$$\sum_{-\infty}^{\infty} |(a+b)(n)| = \sum_{-\infty}^{\infty} |a(n) + b(n)| \leq \sum_{-\infty}^{\infty} \{|a(n)| + |b(n)|\}$$

$$= \sum_{-\infty}^{\infty} |a(n)| + \sum_{-\infty}^{\infty} |b(n)|$$

and these last two sums are finite. It is also clear that when $a \epsilon \ell^1(\mathbb{Z})$, so does λa for any scalar λ. One can now check that our set has all of the properties listed in Definition 1 of Appendix A, and so $\ell^1(\mathbb{Z})$ is a vector space.

For each $a \epsilon \ell^1(\mathbb{Z})$ we define $\|a\|_1$ to be

$$\sum_{-\infty}^{\infty} |a(n)|$$

This is a nonnegative real-valued function and $\|a\|_1 = 0$ if, and only if, $a(n) = 0$ for all $n \epsilon Z$; i.e., if, and only if, a is the zero vector in $\ell^1(\mathbb{Z})$. Also, $\|\lambda a\|_1 = |\lambda| \|a\|_1$ for any a and any scalar λ. The triangle inequality (property (a) in Definition 1) was proved in the paragraph just above. Thus $(\ell^1(\mathbb{Z}), \|\cdot\|_1)$ is a normed space.

We shall see more examples of normed spaces in the later sections.

Definition 2. Let $(E, \|\cdot\|)$ be a normed space and let $\vec{x}_0 \epsilon E$. A subset S of E is said to be a neighborhood of $\vec{x}_0$ if for some $\varepsilon > 0$ we have $\{\vec{x} \epsilon E \mid \|\vec{x} - \vec{x}_0\| < \varepsilon\} \subseteq S$. A subset of E is said to be an open set if it is a neighborhood of each of its points, and a subset of E is said to be a closed set if its compliment, in E, is an open set.

Given $\vec{x}_0 \epsilon E$ and any $\varepsilon > 0$, the set $O_\varepsilon(\vec{x}_0) = \{\vec{x} \epsilon E \mid \|\vec{x} - \vec{x}_0\| < \varepsilon\}$ is called the open ball of radius ε centered at $\vec{x}_0$, and the set $\beta_\varepsilon(\vec{x}_0) = \{\vec{x} \epsilon E \mid \|\vec{x} - \vec{x}_0\| \leq \varepsilon\}$ is called the closed ball of radius ε centered at $\vec{x}_0$. Observe that an open ball is an open set and a closed ball is a closed set (see Exercises 1, problem 3). For any $A \subseteq E$ and any scalar λ, let $\lambda A = \{\lambda \vec{x} \mid \vec{x} \epsilon A\}$. We have

$$\beta_\varepsilon\left(\vec{0}\right) = \{\vec{x} \epsilon E \mid \|\vec{x}\| \leq \varepsilon\} = \{(\varepsilon \vec{x}) \mid \|\vec{x}\| \leq 1\} = \varepsilon \beta_1(\vec{0}) \tag{1}$$

For any set $B \subseteq E$ let $A + B = \{\vec{x} + \vec{y} \mid \vec{x} \epsilon A \text{ and } \vec{y} \epsilon B\}$
Then

$$\beta_\varepsilon\left(\vec{x}_0\right) = \left\{\vec{x} \epsilon E \mid \|\vec{x} - \vec{x}_0\| \leq \varepsilon\right\} \tag{2}$$

$$= \left\{\vec{y} \epsilon E \mid \vec{y} = \vec{x}_0 + \varepsilon \vec{x} \text{ with } \|\vec{x}\| \leq 1\right\} = \left\{\vec{x}_0\right\} + \varepsilon \beta_1(\vec{0})$$

because

$$\{\vec{x}_0\} + \varepsilon\beta_1(\vec{0}) = \{\vec{x}_0 + \varepsilon\vec{x} \mid \|\vec{x}\| \leq 1\} \qquad (3)$$

Hence we see that if we call $\beta_1(\vec{0})$ the unit ball of E, any closed ball centered at zero is a scalar times the unit ball and any closed ball with center $\vec{x}_0$ is just a translate to $\vec{x}_0$ of a closed ball with center at zero.

Similar remarks apply to open balls.

Exercises 1

* 1. Let $(E, \|\cdot\|)$ be a normed space and let $\vec{x}, \vec{y}$ be elements of E. Show that $|\,\|\vec{x}\| - \|\vec{y}\|\,| \leq \|\vec{x} - \vec{y}\|$.

    Hint: Write $\vec{x} = \vec{x} - \vec{y} + \vec{y}$ and $\vec{y} = \vec{y} - \vec{x} + \vec{x}$.

* 2. Find a vector $(z_1, z_2) \in \mathbb{C}^2$ such that $z_1^2 + z_2^2 = 0$ and $|z_1|^2 + |z_2|^2 \neq 0$.

* 3. (a) Show that an open ball, say $O_\varepsilon(\vec{x}_0)$, is an open set. Given $\vec{y} \in O_\varepsilon(\vec{x}_0)$, find $\delta > 0$ such that $O_\delta(\vec{y}) \subseteq O_\varepsilon(\vec{x}_0)$.

    (b) Show that a closed ball, say $\beta_\varepsilon(\vec{x}_0)$, is a closed set. Given $\vec{y} \notin \beta_\varepsilon(\vec{x}_0)$, find $\delta > 0$ such that $O_\delta(\vec{y}) \cap \beta_\varepsilon(\vec{x}_0) = \emptyset$.

    (c) Show that the union of any collection of open sets is an open set, and the intersection of any finite collection of open sets is an open set.

    (d) Show that the intersection of any family of closed sets is a closed set, and that the union of any finite collection of closed sets is a closed set.

4. In $\mathbb{R}^n$ the vectors $\vec{e}_1 = (1, 0, 0, ..., 0)$, $\vec{e}_2 = (0, 1, 0, ..., 0)$, $\vec{e}_3 = (0, 0, 1, 0, ..., 0), ..., \vec{e}_n = (0, ..., 0, 1)$ are linearly independent.

    An infinite subset of a vector space is said to be linearly independent if each of its finite subsets is linearly independent. Find an infinite linearly independent subset of $\ell^2(\mathbb{N})$ and find such a set in $\ell^1(\mathbb{Z})$. What does this tell you about the dimension of these vector spaces?

2. Convergence in Normed Spaces

The presence of a norm on a vector space enables us to define convergence for sequences in that space. Once we have this concept, we shall be able to discuss convergent series in, and continuous functions on, the space.

Definition 1. Let $(E, \|\cdot\|)$ be a normed space and let $\{\vec{x}_n\}_{n=1}^{\infty}$ be a sequence of vectors in E. We shall say that this sequence is convergent to the vector $\vec{x}_0 \epsilon E$, and we shall write $\lim_{n \to \infty} \vec{x}_n = \vec{x}_0$, if for any given $\varepsilon > 0$ there is a natural number N such that $\|\vec{x}_n - \vec{x}_0\| < \varepsilon$ whenever $n \geq N$.

Note that the sequence $\{\vec{x}_n\} \subseteq E$ converges to $\vec{x}_0 \epsilon E$ if, and only if, the sequence $\{\vec{x}_n - \vec{x}_0\}$ converges to the zero vector of E. As a simple application of this idea, let us use convergent sequences to characterize the closed subsets of $(E, \|\cdot\|)$.

Lemma 1. A nonempty subset C of E is closed in $(E, \|\cdot\|)$ if, and only if, every sequence of points of C that is convergent converges to a point in C.

Proof. Suppose first that C is closed and that $\{\vec{x}_n\} \subseteq C$ is convergent to a point $\vec{y} \epsilon E$. If $\vec{y} \notin C$, then $\vec{y}$ is in the complement of C, $E \setminus C$, which is an open set. Hence, for some $\epsilon > 0$ we must have $O_\epsilon(\vec{y}) \subseteq E \setminus C$ (Section 1, Definition 2). Because $\{\vec{x}_n\}$ converges to $\vec{y}$, there must be an integer N such that $\vec{x}_n \epsilon O_\varepsilon(\vec{y})$ whenever $n \geq N$. Thus we have reached a contradiction because, when $n \geq N$, $\vec{x}_n$ is in both C and $E \setminus C$.

Now suppose that C is **not** a closed set. We shall construct a sequence $\{\vec{x}_n\} \subseteq C$ that converges to a point $\vec{y}$ that is not in C. Because C is not closed, $E \setminus C$ is not open, and hence, in particular, it cannot be empty. Thus there must be a point $\vec{y} \epsilon E \setminus C$ such that every neighborhood of $\vec{y}$ intersects C; otherwise $E \setminus C$ would be a neighborhood of each of its points. So for each $n = 1, 2, ...$ we can find $\vec{x}_n \epsilon O_{\frac{1}{n}}(\vec{y}) \cap C$. Clearly, $\{\vec{x}_n\}$ is a sequence in C whose limit is $\vec{y} \notin C$.

The Fourier transform is of fundamental importance in harmonic analysis. It is a linear map between two normed spaces. Whenever we speak of a linear map between normed spaces E_1, E_2 we shall always assume that these spaces are defined over the same field $\mathbb{K}$ (i.e., both are vector spaces over $\mathbb{R}$ or both are vector spaces over $\mathbb{C}$). Recall that a map $T: E_1 \to E_2$ is said to be linear if

(i) $T(\vec{x} + \vec{y}) = T(\vec{x}) + T(\vec{y})$ for all $\vec{x}, \vec{y}$ in E_1 and

(ii) $T(\lambda \vec{x}) = \lambda T(\vec{x})$ for all $\vec{x} \epsilon E_1$ and all scalars λ.

Observe that T must map the zero vector in E_1 onto the zero vector in E_2.

Definition 2. Let $(E_1, \|\cdot\|_1)$ and $(E_2, \|\cdot\|_2)$ be two normed spaces and let $T: E_1 \to E_2$ be a linear map. We shall say that T is a continuous linear map if the sequence $\{T(\vec{x}_n)\}$ converges to the vector $T(\vec{x}_0)$ for $\|\cdot\|_2$ whenever the sequence $\{\vec{x}_n\} \subseteq E_1$ converges to the vector $\vec{x}_0$ for $\|\cdot\|_1$.

Our first result gives some especially useful characterizations of continuous linear maps. It is because of the first two of these that continuous linear maps are often called "bounded" linear mappings.

Theorem 1. Let $(E_1, \|\cdot\|_1)$ and $(E_2, \|\cdot\|_2)$ be two normed spaces and let $T: E_1 \to E_2$ be a linear mapping. Then the following are equivalent:

(i) There is a real number M such that $\|T(\vec{x})\|_2 \leq M\|\vec{x}\|_1$ for all $\vec{x}\epsilon E_1$;

(ii) There is a real number M such that $\|T(\vec{x})\|_2 \leq M$ whenever $\vec{x}\epsilon E_1$ and $\|x\|_1 \leq 1$;

(iii) T is a continuous linear map.

Proof. It is clear that (i) implies (ii). Suppose that T has property (ii). Then given $\vec{x}\epsilon E_1$, $\vec{x} \neq \vec{0}$, we must have $\|T\left(\frac{\vec{x}}{\|\vec{x}\|_1}\right)\|_2 \leq M$. Because T is linear we see that $\|T(\vec{x})\|_2 \leq M\|\vec{x}\|_1$ for any nonzero vector $\vec{x}\epsilon E_1$. But any linear map takes the zero vector to the zero vector, so our inequality holds when $\vec{x}$ is $\vec{0}$ as well. Thus (i) and (ii) are equivalent.

Let us now show that (i) implies (iii). If $\{\vec{x}_n\}$ is any sequence in E_1 that converges to a vector $\vec{x}_0\epsilon E_1$, for $\|\cdot\|_1$, then $\{\vec{x}_0 - \vec{x}_n\}$ converges to the zero vector of E_1 for $\|\cdot\|_1$. But by (i) we must have $\|T(\vec{x}_0 - \vec{x}_n)\|_2 \leq M\|(\vec{x}_0 - \vec{x}_n)\|_1$, and so it follows from the linearity of T that $\lim_{n\to\infty} T(\vec{x}_n) = T(\vec{x}_0)$ for all $\|\cdot\|_2$.

Finally, let us show that (iii) implies (ii). We shall assume (iii) and the negation of (ii) and deduce a contradiction: We are supposing that, given any M, we can find $\vec{x}\epsilon E_1$ such that $\|\vec{x}\|_1 \leq 1$ and $\|T(\vec{x})\|_2 > M$. Choose $\vec{x}_1\epsilon E_1$ so that $\|\vec{x}_1\|_1 \leq 1$ and $\|T(\vec{x}_1)\|_2 > 1^2$. Next choose $\vec{x}_2\epsilon E_1$ so that $\|\vec{x}_2\|_1 \leq 1$ and $\|T(\vec{x}_2)\|_2 > 2^2$. Continue in this way. After $\vec{x}_1, \vec{x}_2, ... \vec{x}_n$ have been chosen so that $\|\vec{x}_j\|_1 \leq 1$ and $\|T(\vec{x}_j)\|_2 > j^2$ for $j = 1, 2, ..., n$, choose $\vec{x}_{n+1}\epsilon E_1$ so that $\|\vec{x}_{n+1}\|_1 \leq 1$ and $\|T(\vec{x}_{n+1})\|_2 > (n+1)^2$.

Now the sequence $\left\{\frac{\vec{x}_n}{n}\right\} \subseteq E_1$ satisfies the inequalities $\|\frac{\vec{x}_n}{n}\|_1 \leq \frac{1}{n}$ and $\|T\left(\frac{\vec{x}_n}{n}\right)\|_2 > n$ for all n. But then the sequence $\left\{\frac{\vec{x}_n}{n}\right\} \subseteq E_1$ converges to the zero vector of E_1 for $\|\cdot\|_1$, whereas the sequence $\left\{T\left(\frac{\vec{x}_n}{n}\right)\right\} \subseteq E_2$ does not converge. This contradicts (iii) and completes the proof.

A linear map from a vector space to itself is called a linear operator on that space. We have two natural linear operators on the space $\ell^2(\mathbb{N})$. They are the right-shift operator

$$A_r\left(\vec{z}\right) = A_r(\{z_n\}_{n=1}^{\infty}) = \{0, z_1, z_2, ...\} \quad \text{for all } \vec{z} \in \ell^2(\mathbb{N})$$

and the left-shift operator

$$A_\ell\left(\vec{z}\right) = A_\ell(\{z_n\}_{n=1}^{\infty}) = \{z_n\}_{n=2}^{\infty} \quad \text{for all } \vec{z} \in \ell^2(\mathbb{N})$$

Clearly, $\|A_r\left(\vec{z}\right)\|_2 = \|\vec{z}\|_2$ for every $\vec{z}$, and $\|A_\ell\left(\vec{z}\right)\|_2 \leq \|\vec{z}\|_2$ for every $\vec{z}$. Thus these linear operators are continuous on $\ell^2(\mathbb{N})$.

Because A_r and A_ℓ map $\ell^2(\mathbb{N})$ to itself, we can compose them (Chapter Zero, Exercises 2, problem 2). We see that $A_\ell \circ A_r$ is the identity operator (i.e., the map I such that $I\vec{z} = \vec{z}$ for all $\vec{z}$) whereas $A_r \circ A_\ell$ is not.

The space $\ell^1(Z)$ consists of all sequences $\vec{z} = \{z_n\}_{-\infty}^{\infty}$ such that

$$\sum_{-\infty}^{\infty} |z_n| < \infty$$

This last number is $\|\vec{z}\|_1$. Clearly,

$$\sum_{-\infty}^{\infty} z_n$$

is convergent and so we may define a map $\varphi \colon \ell^1(z) \to \mathbb{C}$ by setting

$$\varphi(\vec{z}) = \varphi(\{z_n\}) = \sum_{-\infty}^{\infty} z_n$$

for every $\vec{z} \in \ell^1(Z)$. This mapping is easily seen to be linear. A linear map from a vector space into its field of scalars is called a linear functional on the space. Because $\mathbb{K}$ has a norm (the absolute value function [see Chapter Zero, Exercises

3, problem 3 and Section 4], is a norm on $\mathbb{K}$), we may consider the continuous linear functionals on a normed space. Note that φ is continuous because

$$|\varphi(\vec{z})| = |\varphi(\{z_n\})| = \left|\sum_{-\infty}^{\infty} z_n\right| \leq \sum_{-\infty}^{\infty} |z_n| = \|\vec{z}\|_1$$

for every $\vec{z} \in \ell^1(Z)$.

Now let us consider a fixed nonempty subset D of $\mathbb{K}$ and a function $f : D \to \mathbb{K}$. We shall say that f is a bounded function or that f is bounded on D, if there is a real number M such that $|f(t)| \leq M$ for all $t \epsilon D$. Such a function need not have a maximum in D but the least upper bound of the set $\{|f(t)| \| t \epsilon D\}$, what we call the supremium of $|f|$, is a real number (Chapter Zero, Section 3, Definition 1 and Remark (8)).

Now let $B(D)$ be the set of all functions from D into $\mathbb{K}$ that are bounded on D. This is a vector space over $\mathbb{K}$ (Appendix A) and we can define a norm on this space by setting $\|f\|_\infty = \sup |f| = lub\{|f(t)| \| t \epsilon D\}$ (we call this the sup norm) for each $f \epsilon B(D)$. Furthermore, we can easily characterize the sequences of functions that converge for this norm. We have:

Lemma 2. A sequence $\{f_n\} \subseteq B(D)$ converges to the function $f_0 \epsilon B(D)$ for the norm $\|\cdot\|_\infty$ if, and only if, this sequence converges to f_0 uniformly over D.

Proof. Let $\{f_n\} \subseteq B(D)$ and suppose that this sequence converges to $f_0 \epsilon B(D)$ for $\|\cdot\|_\infty$. Then, given $\varepsilon > 0$, there is an integer N such that $\|f_0 - f_n\|_\infty < \varepsilon$ whenever $n \geq N$. But this says $|f_0(t) - f_n(t)| \leq lub\{|f_0(t) - f_n(t)| \| t \epsilon D\} < \varepsilon$, for every $t \epsilon D$, when $n \geq N$. Hence, $\{f_n\}$ converges to f_0 uniformly over D.

Now suppose that the sequence $\{f_n\}$ converges to f_0 uniformly over D. Then given $\varepsilon > 0$ there is an N such that $|f_0(t) - f_n(t)| < \varepsilon$ for all $t \epsilon D$ whenever $n \geq N$. Clearly then, $\|f_0 - f_n\|_\infty = lub\{|f_0(t) - f_n(t)| \| t \epsilon D\} \leq \varepsilon$ whenever $n \geq N$ showing $\{f_n\}$ converges to f_0 for $\|\cdot\|_\infty$.

In Chapter One, Section 4, Lemma 1, we showed that a sequence $\{f_n\}$ of functions defined on D is uniformly convergent over D if given $\varepsilon > 0$ there is an N such that $|f_n(t) - f_m(t)| < \varepsilon$ for all $t \epsilon D$ whenever both m and n exceed N. It follows that a sequence $\{f_n\} \subseteq B(D)$ converges for $\|\cdot\|_\infty$ if, and only if, given $\varepsilon > 0$ there is an N such that $\|f_n - f_m\|_\infty < \varepsilon$ whenever both m and n exceed N. Normed spaces that have this property are extremely important in both pure and applied mathematics. Before we can discuss them we must define our terms.

Definition 3. Let $(E, \|\cdot\|)$ be a normed space. A sequence $\{\vec{x}_n\}$ of vectors in E is said to be a Cauchy sequence if, given $\varepsilon > 0$, there is an N such that

$\|\vec{x}_n - \vec{x}_m\| < \varepsilon$ whenever m and n both exceed N. A normed space $(E, \|\cdot\|)$ is said to be a complete normed space, or a Banach space, if every Cauchy sequence of points of E converges to a point of E.

We have just noted that the space $(B(D), \|\cdot\|_\infty)$ is a Banach space. In particular, $(B(T), \|\cdot\|_\infty)$, T the unit circle, is a Banach space. Observe that the continuous functions on T are bounded and hence $C(T) \subseteq B(T)$. Furthermore, any sequence in $C(T)$ that is a Cauchy sequence must have its limit in $C(T)$. Because it is Cauchy it converges to an element of $B(T)$ and because each f_n is continuous, its limit must actually be in $C(T)$ (Chapter One, Section 4, Lemma 2). Thus $(C(T), \|\cdot\|_\infty)$ is another example of a complete normed space.

Exercises 2

*1. Let $(E, \|\cdot\|)$ be a normed space and let $\{\vec{x}_n\} \subseteq E$ converge to $\vec{x}_0 \epsilon E$ for $\|\cdot\|$.

  (a) Show that the sequence of real numbers $\{\|\vec{x}_n\|\}$ converges to the number $\|\vec{x}_0\|$.

  (b) Show that $\{\vec{x}_n\}$ is a Cauchy sequence (Definition 3).

*2. Let S be a nonempty subset of a normed space $(E, \|\cdot\|)$. A point $\vec{x}_0$ of E is said to be an adherent point of S if every neighborhood of $\vec{x}_0$ contains infinitely many points of S. The set of all adherent points of S is denoted by S', and the set $S \cup S'$, which we shall call the closure of S, is denoted by $\bar{S}$.

  (a) Show that $\vec{x}_0 \epsilon S'$ if, and only if, there is a sequence of distinct points of S that converges to $\vec{x}_0$.

  (b) Show that $(S')' \subseteq S'$ and hence S' is a closed set.

  (c) Show that $\bar{S}$ is a closed set and show that any closed set that contains S must also contain $\bar{S}$.

  (d) If H is a linear subspace of E, show that $\bar{H}$ is also a linear subspace of E. A linear subspace of E that is also a closed set is called a closed linear subspace of E.

  (e) Show that $C(T)$ is a closed linear subspace of $(B(T), \|\cdot\|)$.

3. For each $n \epsilon \mathbb{N}$ let $f_{2n}(t) = t^{2n}$ for $-\pi \leq t \leq \pi$ and let $f_{2n}(t + 2\pi) = f_{2n}(t)$ for all t. Then $\{f_{2n}\}_{n=1}^\infty \subseteq C(T)$. Show that any finite subset of $\{f_{2n}\}$ is linearly independent.

4. Let E_1, E_2 be two normed spaces and let $T : E_1 \to E_2$ be a continuous linear map.

   (a) Show that the null space (Appendix A) of T is a closed linear subspace of E_1.

   (b) More generally, show that for any closed subset C of E_2 the set $T^{-1}(C) = \{\vec{x} \epsilon E_1 \mid T(\vec{x}) \epsilon C\}$ is a closed subset of E_1.

5. We can define two linear functionals on $C(T)$ as follows:

   (a) $I(f) = \int_{-\pi}^{\pi} f(t)\,dt$ for each $f \epsilon C(T)$;

   (b) For any fixed $s \in T$, let $\varphi_s(f) = f(s)$ for each $f \epsilon C(T)$. Show that these maps are continuous when $C(T)$ has the sup norm.

6. Recall the space $\ell^1(Z)$ (Section 1, example (iv)). For each $a \epsilon \ell^1(Z)$ let
$$\mathcal{F}(a) = \sum_{-\infty}^{\infty} a(n) e^{in\theta}.$$

   (a) Show that $\mathcal{F}(a) \epsilon C(T)$ and show that the map taking each a to $\mathcal{F}(a)$, call it $\mathcal{F}$, is linear.

   (b) Show that $\mathcal{F}$ is a continuous linear mapping from $(\ell^1(Z), \|\cdot\|_1)$ into $(C(T), \|\cdot\|_\infty)$.

3. Inner Product Spaces

Returning once again to elementary physics and our informal discussion, we recall that the dot product of two vectors $\vec{u}, \vec{v}$ in $\mathbb{R}^2$, or $\mathbb{R}^3$, is the product of their lengths and the cosine of the angle between these vectors. Thus $\langle \vec{u}, \vec{v} \rangle = \|\vec{u}\|\|\vec{v}\|\cos\theta$. Because we have an inner product on $\mathbb{R}^n$ (introduction to this chapter, equation (1)) and we have a norm on $\mathbb{R}^n$ (Section 1, example (i)), we might define the angle between $\vec{u}, \vec{v} \epsilon \mathbb{R}^n$ to be $\cos^{-1}\left\{\frac{\langle \vec{u}, \vec{v} \rangle}{\|\vec{u}\|\|\vec{v}\|}\right\}$. To carry this out we'd have to show that the quantity in curly brackets is between -1 and $+1$; otherwise the arccosine wouldn't be real. This is, as we shall see, actually the case. So we can give "direction" to the vectors in $\mathbb{R}^n$. But this is not, as far as I know, very useful. What is important about an inner product is that it enables us to define "orthogonality" of vectors. In $\mathbb{R}^2$, or $\mathbb{R}^3$, two non-zero vectors are orthogonal if the angle between them is $\frac{\pi}{2}$. In this case the cosine is zero, so the inner product of the two vectors vanishes. In the general case when we have an inner product defined on the vectors of a given space, we define two vectors to be orthogonal if their inner product is zero. This is an extremely useful notion as we shall see below. Let us now give formal definitions of these concepts.

3. Inner Product Spaces

Definition 1. Let V be a vector space over $\mathbb{K}$. The set of all ordered pairs $\left(\vec{u}, \vec{v}\right)$ whose members are in V is denoted by $V \times V$ (Chapter Zero, Section 1, definition 5). A function φ from $V \times V$ into $\mathbb{K}$ is said to be an inner product on V if

(a) For any $\vec{u}_1, \vec{u}_2, \vec{v}$ in V and any scalars α, β we have $\varphi\left(\alpha \vec{u}_1 + \beta \vec{u}_2, \vec{v}\right) = \alpha \varphi\left(\vec{u}_1, \vec{v}\right) + \beta \varphi\left(\vec{u}_2, \vec{v}\right)$;

(b) For any $\vec{u}, \vec{v}$ in V, $\varphi\left(\vec{u}, \vec{v}\right) = \overline{\varphi\left(\vec{v}, \vec{u}\right)}$, where the bar denotes complex conjugation (Chapter Zero, Exercises 4, problem 1);

(c) For all $\vec{v} \epsilon V$, $\varphi\left(\vec{v}, \vec{v}\right) \geq 0$ with equality holding when, and only when, $\vec{v}$ is the zero vector.

When φ is an inner product on V, it is customary to write $\left\langle \vec{u}, \vec{v} \right\rangle$ in place of $\varphi\left(\vec{u}, \vec{v}\right)$ and to call the pair $(V \langle,\rangle)$ an inner product space.

For the remainder of this section we shall suppose that any vectors we consider are members of a fixed inner product space $(V. \langle,\rangle)$.

It is convenient to define for each $\vec{v} \epsilon V$, $\|\vec{v}\|$ to be $\left(\left\langle \vec{v}, \vec{v} \right\rangle\right)^{\frac{1}{2}}$ and to call this the norm of $\vec{v}$. It is clear that $\|\alpha \vec{v}\| = |\alpha| \|\vec{v}\|$ for any $\vec{v} \epsilon V$ and any $\alpha \epsilon \mathbb{K}$, and it is clear that $\|\vec{v}\| = 0$ if, and only if, $\vec{v}$ is the zero vector. To prove that $\|\cdot\|$ is a norm, we must also show that it satisfies the triangle inequality (Defnition 1(a) of Section 1); i.e., we must show that $\|\vec{u} + \vec{v}\| \leq \|\vec{u}\| + \|\vec{v}\|$ for all $\vec{u}, \vec{v}$ in V. This will follow from Theorem 2 below (Corollary 1).

A vector $\vec{v} \epsilon V$ is called a unit vector when $\|\vec{v}\| = 1$. Such a vector is sometimes said to be normalized. We can normalize any nonzero vector $\vec{v}$ by multiplying it by the scalar $\frac{1}{\|\vec{v}\|}$. A set of vectors is said to be normalized if each vector in the set is a unit vector.

Definition 2. Two vectors $\vec{u}, \vec{v}$ in V are said to be orthogonal if $\left\langle \vec{u}, \vec{v} \right\rangle = 0$. A nonempty set of vectors is said to be an orthogonal set if any two distinct vectors in this set are orthogonal. Finally, an orthogonal set that is also normalized is called an orthonormal set.

It is easy to check that the functions defined by equations (1) and (2) in the introduction to this chapter are inner products on $\mathbb{R}^n$ and $\mathbb{C}^n$, respectively. If, for each $j = 1, 2, ..., n$, we let $\vec{e}_j$ be the vector $(0, ..., 1, 0, ..., 0)$, where the 1 is the jth place, then $\left\{\vec{e}_j\right\}_{j=1}^{n}$ is an orthonormal subset of $\mathbb{R}^n$ and also of $\mathbb{C}^n$.

Theorem 1. (Bessel's inequality, weak form). Let $\{\vec{v}_j\}_{j=1}^m$ be a finite orthonormal set in V. Then for any vector $\vec{v} \in V$ we have $\sum_{j=1}^m |\langle \vec{v}, \vec{v}_j \rangle|^2 \leq \|\vec{v}\|^2$.

Proof. The proof is by direct calculation using the properties listed in Definitions 1 and 2. It is convenient to set $c_j = \langle \vec{v}, \vec{v}_j \rangle$ for $j = 1, 2, ..., m$, and to use different indices in our sums. We have

$$0 \leq \|\vec{v} - \sum_{j=1}^m c_j \vec{v}_j\|^2 = \langle \vec{v} - \sum_{j=1}^m c_j \vec{v}_j, \vec{v} - \sum_{k=1}^m c_k \vec{v}_k \rangle =$$

$$\|\vec{v}\|^2 - \sum_{j=1}^m c_j \langle \vec{v}_j, \vec{v} \rangle - \sum_{k=1}^m \bar{c}_k \langle \vec{v}, \vec{v}_k \rangle + \sum_{j,k=1}^m c_j \bar{c}_k \langle \vec{v}_j, \vec{v}_k \rangle =$$

$$\|\vec{v}\|^2 - \sum_{j=1}^m |c_j|^2 - \sum_{k=1}^m |c_k|^2 + \sum_{j=1}^m |c_j|^2$$

because $\langle \vec{v}_j, \vec{v}_k \rangle = 0$ unless $j = k$. Thus we have

$$0 \leq \|\vec{v}\|^2 - \sum_{j=1}^m |c_j|^2$$

and we are done.

Theorem 2. (Cauchy, Schwarz, Bunyakowski). For any two vectors $\vec{u}, \vec{v}$ in V we have $|\langle \vec{u}, \vec{v} \rangle| \leq \|\vec{u}\| \|\vec{v}\|$.

Proof. The result is trivial when $\vec{v}$ is the zero vector (see Exercises 3, problem 1a). When this is not the case, $\{\frac{\vec{v}}{\|\vec{v}\|}\}$ is a finite orthonormal set. Thus, by Theorem 1 we have $|\langle \vec{u}, \frac{\vec{v}}{\|\vec{v}\|} \rangle|^2 \leq \|\vec{u}\|^2$, giving us $|\langle \vec{u}, \vec{v} \rangle|^2 \leq \|\vec{u}\|^2 \|\vec{v}\|^2$. Taking square roots we have our result.

Remark 1. In our discussion of $\mathbb{R}^n$ we suggested that one might be able to define the angle between two nonzero vectors $\vec{u}, \vec{v}$ by using $\cos^{-1}\left\{\frac{\langle \vec{u}, \vec{v} \rangle}{\|\vec{u}\| \|\vec{v}\|}\right\}$. Theorem 2 shows that when the inner product is real, the quantity in curly brackets is always between -1 and 1. Thus, in this case, the arccosine is always a real number.

Corollary 1. For any two vectors $\vec{u}, \vec{v}$ in V we have $\|\vec{u} + \vec{v}\| \leq \|\vec{u}\| + \|\vec{v}\|$.

Proof. First note that $\langle \vec{u}, \vec{v}\rangle + \langle \vec{v}, \vec{u}\rangle$ is real and less than $2\,|\langle \vec{u}, \vec{v}\rangle|$. Hence we may write

$$\|\vec{u}+\vec{v}\|^2 = \langle \vec{u}+\vec{v}, \vec{u}+\vec{v}\rangle = \langle \vec{u}, \vec{u}+\vec{v}\rangle + \langle \vec{v}, \vec{u}+\vec{v}\rangle = \langle \vec{u},\vec{u}\rangle + [\langle \vec{u},\vec{v}\rangle + \langle \vec{v},\vec{u}\rangle] + \langle \vec{v},\vec{v}\rangle \le \|\vec{u}\|^2 + 2\,|\langle \vec{u},\vec{v}\rangle| + \|\vec{v}\|^2$$

But by Theorem 2 this last quantity is $\le \|\vec{u}\|^2 + 2\|\vec{u}\|\|\vec{v}\| + \|\vec{v}\|^2 = \left(\|\vec{u}\| + \|\vec{v}\|\right)^2$. Taking square roots we have our result.

Remark 2. We have now shown that when $(V \langle,\rangle)$ is an inner product space the function $\vec{v} \to \left(\langle \vec{v}, \vec{v}\rangle\right)^{\frac{1}{2}}$ is a norm on V, i.e., setting $\|\vec{v}\| = \left(\langle \vec{v}, \vec{v}\rangle\right)^{\frac{1}{2}}$ for every $\vec{v} \epsilon V$, $(V, \|\cdot\|)$ is a normed space. We shall call this the norm defined by, or associated with, the inner product. Unless the contrary is explicitly stated, whenever we speak of a norm on an inner product space we shall mean the norm defined by the inner product.

We have now shown, in particular, that the functions defined on $\mathbb{R}^n$ and $\mathbb{C}^n$ in examples (i) and (iii) of Section 1 are norms on these spaces. They are the norms defined by the inner products mentioned in the introduction to this chapter (equations (1) and (2)).

Remark 3. An inner product space $(V, \langle,\rangle)$ always has, as we have just seen, a norm. Hence we can discuss open sets, closed sets, and convergent sequences in the space as well as continuous mappings on the space.

Let us now recall the set $\ell^2_{\mathbb{R}}(\mathbb{N})$ (Section 1, example (ii)). This consists of all sequences $\{x_n\}_{n=1}^{\infty}$ of real numbers such that $\sum_{n=1}^{\infty} x_n^2$ is convergent; i.e., it is the set of all functions $x: \mathbb{N} \to \mathbb{R}$ such that $\sum_{n=1}^{\infty} x(n)^2$ is convergent. We know how to add two such functions and how to multiply any such function by a scalar, but are the results still in $\ell^2_{\mathbb{R}}(\mathbb{N})$? Clearly, $\lambda\{x_n\}_{n=1}^{\infty} = \{\lambda x_n\}_{n=1}^{\infty} \epsilon \ell^2_{\mathbb{R}}(\mathbb{N})$ whenever $\{x_n\}_{n=1}^{\infty}$ is in this set and $\lambda \epsilon \mathbb{R}$. Let us show that for any $\{x_n\}_{n=1}^{\infty}$ and any $\{y_n\}_{n=1}^{\infty}$ in $\ell^2_{\mathbb{R}}(\mathbb{N})$ their sum is again in this set; i.e., let us show that when $\sum_{n=1}^{\infty} x_n^2 < \infty$ and $\sum_{n=1}^{\infty} y_n^2 < \infty$ then $\sum_{n=1}^{\infty} (x_n + y_n)^2$ is finite.

For each fixed N we have $\{x_n\}_{n=1}^{N}$ and $\{y_n\}_{n=1}^{N}$ in the space $\mathbb{R}^N$. This is, as we have seen, an inner product space and so we may use the C.S.B. inequality

(Theorem 2) to obtain

$$\left|\left\langle \{x_n\}_{n=1}^N, \{y_n\}_{n=1}^N \right\rangle\right| = \left|\sum_{n=1}^N x_n y_n\right| \le \left(\sum_{n=1}^N x_n^2\right)^{\frac{1}{2}} \left(\sum_{n=1}^N y_n^2\right)^{\frac{1}{2}}$$

Letting $N \to \infty$ we see that whenever $\{x_n\}$ and $\{y_n\}$ are in $\ell_{\mathbb{R}}^2(\mathbb{N})$ the series $\sum_{n=1}^\infty x_n y_n$ is convergent. Furthermore, because

$$\sum_{n=1}^N (x_n + y_n)^2 = \sum_{n=1}^N x_n^2 + 2\sum_{n=1}^N x_n y_n + \sum_{n=1}^N y_n^2$$

we see, by again letting $N \to \infty$, that $\{x_n + y_n\}_{n=1}^\infty$ is in $\ell_{\mathbb{R}}^2(\mathbb{N})$.

It is easy to check that $\ell_{\mathbb{R}}^2(\mathbb{N})$ is a vector space over $\mathbb{R}$ and if we define for each $\vec{x} = \{x_n\}_{n=1}^\infty$, $\vec{y} = \{y_n\}_{n=1}^\infty$ in this space $<\vec{x}, \vec{y}>$ to be $\sum_{n=1}^\infty x_n y_n$, we easily see that $(\ell_{\mathbb{R}}^2(\mathbb{N}), \langle, \rangle)$ is an inner product space. Thus $\|\cdot\|_2$, where for each $\vec{x} = \{x_n\}_{n=1}^\infty$, $\|\vec{x}\|_2 = \left(\sum_{n=1}^\infty x_n^2\right)^{\frac{1}{2}}$, is a norm.

Exercises 3

In problems 1–3 we assume that all vectors considered are in a fixed inner product space $(V, \langle, \rangle)$.

* 1. (a) Show that $\left\langle \vec{u}, \vec{o} \right\rangle = 0 = \left\langle \vec{o}, \vec{u} \right\rangle$ for any $\vec{u}$.

   (b) Prove the parallelogram identity: For any $\vec{u}, \vec{v}$

   $$\|\vec{u} + \vec{v}\|^2 + \|\vec{u} - \vec{v}\|^2 = 2\left(\|\vec{u}\|^2 + \|\vec{v}\|^2\right)$$

   (c) Prove the Pythagorean relation: if $<\vec{u}, \vec{v}> = 0$, then $\|\vec{u} + \vec{v}\|^2 = \|\vec{u}\|^2 + \|\vec{v}\|^2$.

   (d) Prove the polar identity: For any $\vec{u}, \vec{v}$

   $$\frac{1}{4}\left\{\|\vec{u} + \vec{v}\|^2 - \|\vec{u} - \vec{v}\|^2 + i\|\vec{u} + i\vec{v}\|^2 - i\|\vec{u} - i\vec{v}\|^2\right\} = \left\langle \vec{u}, \vec{v} \right\rangle$$

2. Show that any finite subset of an orthonormal set in V is linearly independent.

*3. Let $\{\vec{u}_n\}$ and $\{\vec{v}_n\}$ be two sequences in V that converge, for the norm of V, to $\vec{u}_0$ and $\vec{v}_0$, respectively.

(a) Show that for any $\vec{w} \epsilon V$, $\lim_{n \to \infty} \langle \vec{u}_n, \vec{w} \rangle = \langle \vec{u}_0, \vec{w} \rangle$.

(b) Show that $\lim_{n \to \infty} \langle \vec{u}_n, \vec{v}_n \rangle = \langle \vec{u}_0, \vec{v}_0 \rangle$.

4. On the space $C(T)$ define, for any $f, g, <f, g>$ to be $\int_{-\pi}^{\pi} f(t) \overline{g(t)} dt$.

(a) Show that $\langle, \rangle$ is an inner product on $C(T)$.

(b) Denote the norm associated with this inner product by $\|\cdot\|_2$ so $\|f\|_2 = (\langle f, f \rangle)^{\frac{1}{2}} = \left(\int_{-\pi}^{\pi} |f(t)|^2 dt \right)^{\frac{1}{2}}$. Compute $\|e^{int}\|_2$ and $\|e^{int}\|_\infty$ (Section 2, just after Definition 3) for any fixed $n \epsilon Z$.

(c) Show that $\left\{ \frac{e^{int}}{\sqrt{2\pi}} \mid n \epsilon Z \right\}, \left\{ \frac{\cos nt}{\sqrt{\pi}} \right\}_{n=1}^{\infty}$ and $\left\{ \frac{\sin nt}{\sqrt{\pi}} \right\}_{n=1}^{\infty}$ are orthonormal sets in $(C(T), <,>)$.

5. Find an infinite orthonormal set in $\ell_{\mathbb{R}}^2(\mathbb{N})$.

6. Recall the set $\ell_{\mathbb{C}}^2(\mathbb{N})$ (Section 1, example (iii). This consists of all sequences of complex numbers, $\vec{z} = \{z_n\}$, such that $\sum_{n=1}^{\infty} |z_n|^2 < \infty$.

(a) Show that for any $\{z_n\}, \{w_n\}$ in $\ell_{\mathbb{C}}^2(\mathbb{N})$, $\sum_{n=1}^{\infty} z_n \bar{w}_n$ is convergent.

(b) For any two nonzero complex numbers z, w let $\alpha = \frac{z\bar{w}}{|z\bar{w}|}$. Note that $|\alpha| = 1$ and $\alpha z \bar{w} = |z\bar{w}|$. Hence if $\{z_n\}$ and $\{w_n\}$ are in $\ell_{\mathbb{C}}^2(\mathbb{N})$, so is $\{\alpha_n z_n\}$ where $\alpha_n = \frac{z_n \bar{w}_n}{|z_n \bar{w}_n|}$. Thus we see that $\sum_{n=1}^{\infty} |z_n w_n|$ is convergent.

(c) For any $\vec{z} = \{z_n\}, \vec{w} = \{w_n\}$ in ℓ^2 we have defined $\vec{z} + \vec{w}$ to be $\{z_n + w_n\}$, and for any $\lambda \epsilon \mathbb{C}$ we have defined $\lambda \vec{z}$ to be $\{\lambda z_n\}$. Show that $\vec{z} + \vec{w}$ and $\lambda \vec{z}$ are in $\ell_{\mathbb{C}}^2(\mathbb{N})$ and that $\ell_{\mathbb{C}}^2(\mathbb{N})$ is a vector space over $\mathbb{C}$.

(d) For any $\vec{z} = \{z_n\}, \vec{w} = \{w_n\}$ in $\ell_{\mathbb{C}}^2(\mathbb{N})$ define $\langle \vec{z}, \vec{w} \rangle$ to be $\sum_{n=1}^{\infty} z_n \bar{w}_n$.

Show that $\langle, \rangle$ is an inner product on $\ell_{\mathbb{C}}^2(\mathbb{N})$ and that the norm associated with this inner product is the one defined in Section 1, example (iii).

(e) Find an infinite orthonormal subset of $(\ell_{\mathbb{C}}^2(\mathbb{N}), \langle, \rangle)$.

4. Infinite Orthonormal Sets, Hilbert Space

Let us begin our discussion with a fixed inner product space $(V, \langle,\rangle)$ and an orthornormal sequence (i.e., a sequence that is an orthonormal set) $\{v_j\} \subseteq V$. We have seen several examples of such sequences (Exercises 3, problems 4c, 5, and 6e). Suppose that for some $\vec{u} \epsilon V$ and some scalars $\{\alpha_j\}$ we have

$$\vec{u} = \sum_{j=1}^{\infty} \alpha_j v_j. \tag{1}$$

What we mean, of course, is that $\vec{u} = \lim_{n \to \infty} \sum_{j=1}^{n} \alpha_j \vec{v}_j$ where convergence is for the norm on V defined by its inner product. Then, by problem 3a of Exercises 3, we may write

$$\begin{aligned}\left\langle \vec{u}, \vec{v}_n \right\rangle &= \left\langle \sum_{j=1}^{\infty} \alpha_j \vec{v}_j, \vec{v}_n \right\rangle = \left\langle \lim_{k \to \infty} \sum_{j=1}^{k} \alpha_j \vec{v}_j, \vec{v}_n \right\rangle \\ &= \lim_{k \to \infty} \left\langle \sum_{j=1}^{k} \alpha_j \vec{v}_j, \vec{v}_n \right\rangle \\ &= \lim_{k \to \infty} \sum_{j=1}^{k} \alpha_j \left\langle \vec{v}_j, \vec{v}_n \right\rangle = \alpha_n \quad \text{for each fixed } n.\end{aligned} \tag{2}$$

Thus when (1) is true the scalars are uniquely determined by $\vec{u}$. They are

$$\alpha_j = \left\langle \vec{u}, \vec{v}_j \right\rangle, j = 1, 2, \cdots \tag{3}$$

We can immediately apply our result to the space $C(T)$. Recall that, for all f, g in this space, $\langle f, g \rangle = \int_{-\pi}^{\pi} f(t) \overline{g(t)} \, dt$. Furthermore, the sequence $\left\{ \dfrac{e^{int}}{\sqrt{2\pi}} \mid n \epsilon Z \right\}$ is orthonormal (Exercises 3, problem 4c). Now suppose that, for some $f \epsilon C(T)$,

$$f = \sum_{-\infty}^{\infty} \alpha_n \frac{e^{int}}{\sqrt{2\pi}} \tag{4}$$

Then, as we saw above

$$\alpha_n = \left\langle f, \frac{e^{int}}{\sqrt{2\pi}} \right\rangle = \frac{1}{\sqrt{2\pi}} \int_{-\pi}^{\pi} f(t) e^{-int} dt = \sqrt{2\pi} c_n(f) \quad \text{for every } n. \tag{5}$$

4. Infinite Orthonormal Sets, Hilbert Space

Putting this into (4) we find that

$$f = \sum_{-\infty}^{\infty} \left(\frac{1}{\sqrt{2\pi}} \int_{-\pi}^{\pi} f(t) e^{-int} dt \right) \frac{e^{int}}{\sqrt{2\pi}} = \sum_{-\infty}^{\infty} c_n(f) e^{int} \qquad (6)$$

where, we recall, that $c_n(f) = \frac{1}{2\pi} \int_{-\pi}^{\pi} f(t) e^{-int} dt$ (Chapter One, Section 5, formula 6)

Thus (4) is the complex Fourier series for f.

Returning now to $(V, \langle, \rangle)$ and the orthonormal sequence $\{\vec{v}_j\}$ we can generalize Theorem 1 of Section 3.

Theorem 1. (Bessel's inequality, strong form). For any vector $\vec{u} \epsilon V$ we have

$$\sum_{j=1}^{\infty} |\langle \vec{u}, \vec{v}_j \rangle|^2 \le \|\vec{u}\|^2.$$

Proof. Let us set $\alpha_j = \langle \vec{u}, \vec{v}_j \rangle$ for each $j = 1, 2, \cdots$. Then
$|\alpha_1|^2 \le |\alpha_1|^2 + |\alpha_2|^2 \le \cdots \le \sum_{j=1}^{m} |\alpha_j|^2 \le \cdots$. But, by the theorem we are generalizing (Section 3, Theorem 1) each term of this increasing sequence of real numbers is bounded by $\|\vec{u}\|^2$. Thus $\sum_{j=1}^{\infty} |\langle \vec{u}, \vec{v}_j \rangle|^2$ converges and this sum is bounded by $\|\vec{u}\|$ as well.

Corollary 1. For any $f \epsilon C(T)$ we have $2\pi \left(\sum_{-\infty}^{\infty} |c_n(f)|^2 \right) \le \|f\|_2^2$ and

$$\pi \left(\sum_{n=1}^{\infty} \{|a_n(f)|^2 + |b_n(f)|^2\} \right) \le \|f\|_2^2.$$

Thus far we have discussed what we can say when we know that the series $\sum_{j=1}^{\infty} \langle \vec{u}, \vec{v}_j \rangle \vec{v}_j$ converges to $\vec{u}$. There is a kind of inner product space for which every series formed in this way will converge; i.e., for every vector $\vec{u}$ in this kind of space the series $\sum_{j=1}^{\infty} \langle \vec{u}, \vec{v}_j \rangle \vec{v}_j$ converges.

Definition 1. Given an inner product space $(V, \langle, \rangle)$ we recall that for each $\vec{v} \epsilon V, \|\vec{v}\| = \langle (\vec{v}, \vec{v}) \rangle^{\frac{1}{2}}$. We shall say that $(V, \langle, \rangle)$ is a Hilbert space if $(V, \|\cdot\|)$ is a complete normed space (Section 2, Definition 3).

We may now state the following:

Corollary 2. If $(V, \langle,\rangle)$ is a Hilbert space, then for every vector $\vec{u} \in V$ the series $\sum_{j=1}^{\infty} \langle \vec{u}, \vec{v}_j \rangle \vec{v}_j$ converges to a point in V.

Proof. Given $\vec{u} \in V$ set $\alpha_j = \langle \vec{u}, \vec{v}_j \rangle$ for each $j = 1, 2, \ldots$. Then for any natural numbers m, n with $n < m$ we may write

$$\|\sum_{j=1}^{m} \alpha_j \vec{v}_j - \sum_{j=1}^{n} \alpha_j \vec{v}_j\|^2 = \|\sum_{j=n+1}^{m} \alpha_j \vec{v}_j\|^2$$

$$= \left\langle \sum_{j=n+1}^{m} \alpha_j \vec{v}_j, \sum_{j=n+1}^{m} \alpha_j \vec{v}_j \right\rangle = \sum_{j=n+1}^{m} |\alpha_j|^2 \quad (7)$$

because $\{\vec{v}_j\}$ is an orthonormal set. But by Theorem 1, the series $\sum_{j=1}^{\infty} |\alpha_j|^2$ is convergent. Hence, given $\varepsilon > 0$ we can choose N so that $\sum_{j=n+1}^{m} |\alpha_j|^2 < \varepsilon$ when $m > n \geq N$. It follows from (7) that $\left\{\sum_{j=1}^{n} \alpha_j \vec{v}_j\right\}_{n=1}^{\infty}$ is a Cauchy sequence in $(V, \|\cdot\|)$. Because this is a Hilbert space, every such sequence converges to a point of V.

Unfortunately, the series $\sum_{j=1}^{\infty} \langle \vec{u}, \vec{v}_j \rangle \vec{v}_j$, even when it does converge, does not necessarily converge to $\vec{u}$. To see why this is so, suppose that the series does converge to $\vec{u}$ and suppose, further, that $\langle \vec{u}, \vec{v}_1 \rangle \neq 0$. Then $\{\vec{v}_j\}_{j=2}^{\infty}$ is an orthonormal set, and the series $\sum_{j=2}^{\infty} \langle \vec{u}, \vec{v}_j \rangle \vec{v}_j$ does converge, but not to $\vec{u}$.

To ensure that the series converges to the vector that generates it, we must somehow guarantee that we have not left any vectors out of our orthonormal set. We can do this by requiring that our set has the following property.

Definition 2. Let $(V, \langle,\rangle)$ be an inner product space. An orthonormal subset S of V is said to be a maximal orthonormal set if the conditions (i) S' an orthonormal set and (ii) $S \subseteq S'$, imply that $S = S'$.

We leave it to the reader to show that an orthonormal set S is maximal if $\langle \vec{u}, \vec{v} \rangle = 0$ for all $\vec{v} \in S$ implies $\vec{u} = \vec{0}$ (Exercises 4, problem 1).

The discussion just before Definition 2 has led some to call a maximal orthonormal set a "complete orthonormal set." We shall not use this terminology.

Hilbert spaces first arose in the papers of David Hilbert on the subject of integral equations, and it is well known that he was attracted to this area by the pioneering work of Ivar Fredholm. These spaces play an important role in establishing the foundations of modern quantum mechanics. Their importance for the study of trigonometric series stems from the following result.

Theorem 2. Let $(H, \langle,\rangle)$ be a Hilbert space and let $\{\vec{v}_j\}_{j=1}^{\infty}$ be an orthonormal sequence of vectors in H. If this sequence has any one of the four following properties, it has all four of them:

(a) $\{\vec{v}_j\}_{j=1}^{\infty}$ is a maximal orthonormal set;

(b) For each $\vec{u} \in H$ the series $\sum_{j=1}^{\infty} \langle \vec{u}, \vec{v}_j \rangle \vec{v}_j$ converges to $\vec{u}$;

(c) (Parseval's relation) For each $\vec{u} \in H$ we have $\|\vec{u}\|^2 = \sum_{j=1}^{\infty} |\langle \vec{u}, \vec{v}_j \rangle|^2$;

(d) (also called Parseval's relation) For any two vectors $\vec{u}, \vec{v}$ in H we have
$$\langle \vec{u}, \vec{v} \rangle = \sum_{j=1}^{\infty} \langle \vec{u}, \vec{v}_j \rangle \overline{\langle \vec{v}, \vec{v}_j \rangle}.$$

Proof. Assume that $\{\vec{v}_j\}$ is maximal and let $\vec{u}$ be any vector in H. We have seen, in Corollary 2 above, that $\sum_{j=1}^{\infty} \langle \vec{u}, \vec{v}_j \rangle \vec{v}_j$ converges to some element of H. We must show that its limit is actually the given vector $\vec{u}$. For any fixed n we have

$$\left\langle \vec{u} - \sum_{j=1}^{\infty} \langle \vec{u}, \vec{v}_j \rangle \vec{v}_j, \vec{v}_n \right\rangle = \left\langle \vec{u} - \lim_{m \to \infty} \sum_{j=1}^{m} \langle \vec{u}, \vec{v}_j \rangle \vec{v}_j, \vec{v}_n \right\rangle$$

$$= \left\langle \lim_{m \to \infty} \left(\vec{u} - \sum_{j=1}^{m} \langle \vec{u}, \vec{v}_j \rangle \vec{v}_j \right), \vec{v}_n \right\rangle$$

and by Exercises 3, problem 3a this is

$$= \lim_{m \to \infty} \left\langle \vec{u} - \sum_{j=1}^{m} \left\langle \vec{u}, \vec{v}_j \right\rangle \vec{v}_j, \vec{v}_n \right\rangle$$

$$= \left\langle \vec{u}, \vec{v}_n \right\rangle - \lim_{m \to \infty} \sum_{j=1}^{m} \left\langle \vec{u}, \vec{v}_j \right\rangle \left\langle \vec{v}_j, \vec{v}_n \right\rangle = \left\langle \vec{u}, \vec{v}_n \right\rangle - \left\langle \vec{u}, \vec{v}_n \right\rangle = 0$$

Thus $\vec{u} - \sum \left\langle \vec{u}, \vec{v}_j \right\rangle \vec{v}_j$ must be the zero vector because $\{\vec{v}_j\}_{j=1}^{\infty}$ is a maximal orthonormal set (Exercises 4, problem 1). So we have shown that (a) implies (b). Now suppose our set has property (b). Then for any $\vec{u} \epsilon H$ we may write

$$\|\vec{u}\|^2 = \left\langle \vec{u}, \vec{u} \right\rangle = \left\langle \sum_{j=1}^{\infty} \left\langle \vec{u}, \vec{v}_j \right\rangle \vec{v}_j, \vec{u} \right\rangle = \left\langle \lim_{m \to \infty} \sum_{j=1}^{m} \left\langle \vec{u}, \vec{v}_j \right\rangle \vec{v}_j, \vec{u} \right\rangle$$

and again by Exercises 3, problem 3a, this becomes

$$= \lim_{m \to \infty} \left\langle \sum_{j=1}^{m} \left\langle \vec{u}, \vec{v}_j \right\rangle \vec{v}_j, \vec{u} \right\rangle = \lim_{m \to \infty} \sum_{j=1}^{m} \left\langle \vec{u}, \vec{v}_j \right\rangle \left\langle \vec{v}_j, \vec{u} \right\rangle$$

$$= \lim_{m \to \infty} \sum_{j=1}^{m} \left\langle \vec{u}, \vec{v}_j \right\rangle \overline{\left\langle \vec{u}, \vec{v}_j \right\rangle}$$

$$= \lim_{m \to \infty} \sum_{j=1}^{m} |\left\langle \vec{u}, \vec{v}_j \right\rangle|^2 = \sum_{j=1}^{\infty} |\left\langle \vec{u}, \vec{v}_j \right\rangle|^2$$

Thus (b) implies (c). But for any vector $\vec{u} \epsilon H$ a direct calculation shows that

$$\|\vec{u} - \sum_{j=1}^{m} \left\langle \vec{u}, \vec{v}_j \right\rangle \vec{v}_j\|^2 = \|\vec{u}\|^2 - \sum_{j=1}^{m} |\left\langle \vec{u}, \vec{v}_j \right\rangle|^2$$

It follows from this that (c) implies (b) and hence these two conditions are equivalent.

Suppose now that $\{\vec{v}_j\}$ has property (c) and let $\vec{u}, \vec{v}$ be any two vectors in H. To prove that (c) implies (d), it suffices to show that (b) implies (d). Now

$$\langle \vec{u}, \vec{v} \rangle = \left\langle \vec{u}, \sum_{j=1}^{\infty} \langle \vec{v}, \vec{v}_j \rangle \vec{v}_j \right\rangle$$

$$= \left\langle \vec{u}, \lim_{m \to \infty} \sum_{j=1}^{m} \langle \vec{v}, \vec{v}_j \rangle \vec{v}_j \right\rangle$$

$$= \lim_{m \to \infty} \left\langle \vec{u}, \sum_{j=1}^{m} \langle \vec{v}, \vec{v}_j \rangle \vec{v}_j \right\rangle$$

$$= \lim_{m \to \infty} \sum_{j=1}^{m} \overline{\langle \vec{v}, \vec{v}_j \rangle} \langle \vec{u}, \vec{v}_j \rangle = \sum_{j=1}^{\infty} \langle \vec{u}, \vec{v}_j \rangle \overline{\langle \vec{v}, \vec{v}_j \rangle}$$

And this is (d).

Finally, we show that (d) implies (a). We shall show that assuming our set has property (d), the only vector in H that is orthogonal to every $\vec{v}_j$ is the zero vector. So suppose $\vec{v} \epsilon H$ and $\langle \vec{v}, \vec{v}_j \rangle = 0$ for every j. Then by (d), we have

$$\|\vec{v}\|^2 = \langle \vec{v}, \vec{v} \rangle = \sum_{j=1}^{\infty} \langle \vec{v}, \vec{v}_j \rangle \overline{\langle \vec{v}, \vec{v}_j \rangle}$$

By our own assumption each term of the series is zero showing that $\|\vec{v}\| = 0$ and hence that $\vec{v}$ is the zero vector.

Definition 3. Let $(H, \langle, \rangle)$ be a Hilbert space. An orthonormal sequence that has any one, and hence all four, of the properties listed in Theorem 2 is called an orthonormal basis for H.

Let us show that the inner product space $(\ell_{\mathbb{R}}^2(\mathbb{N}), \langle, \rangle)$ is a Hilbert space. Given a Cauchy sequence $\{\vec{x}_n\}_{n=1}^{\infty}$ in this space we first note that for each fixed n

$$\vec{x}_n = \{x_{n1}, x_{n2}, x_{n3}, \ldots\} \qquad (1)$$

Now given $\varepsilon > 0$ we can find an integer N such that

$$\|\vec{x}_n - \vec{x}_m\|_2 < \varepsilon \qquad (2)$$

when $m, n \geq N$. Using (1) we may rewrite (2) as follows:

$$\sum_{j=1}^{\infty} (x_{nj} - x_{mj})^2 < \varepsilon^2 \qquad (3)$$

for $m, n \geq N$. For any fixed j the sequence $\{x_{nj}\}_{n=1}^{\infty}$ is a Cauchy sequence of real numbers because

$$|x_{nj} - x_{mj}|^2 = (x_{nj} - x_{mj})^2 \leq \sum_{j=1}^{\infty}(x_{nj} - x_{mj})^2 < \varepsilon^2$$

when $m, n \geq N$. Let us set $x_{0j} = \lim_{n \to \infty} x_{nj}$ for each $j = 1, 2, \ldots$. We shall first show that $\{x_{0j}\}_{j=1}^{\infty} \in \ell_{\mathbb{R}}^2(\mathbb{N})$. Choose and fix $n > N$ and note that for any k we have $\{x_{0j}\}_{j=1}^{k}$ and $\{x_{nj}\}_{j=1}^{k}$ in the space $\mathbb{R}^k$. Because $\|\cdot\|_2$ is a norm on this space we may use the triangle inequality to write

$$\left(\sum_{j=1}^{k} x_{0j}^2\right)^{\frac{1}{2}} \leq \left(\sum_{j=1}^{k}(x_{0j} - x_{nj})^2\right)^{\frac{1}{2}} + \left(\sum_{j=1}^{k} x_{nj}^2\right)^{\frac{1}{2}} \quad (4)$$

Letting $k \to \infty$ on the right-hand side of (4) we have, by (3),

$$\left(\sum_{j=1}^{k} x_{0j}^2\right)^{\frac{1}{2}} \leq \varepsilon + \|\vec{x}_n\|_2$$

Letting $k \to \infty$ we see that $\{\vec{x}_0\} \in \ell_{\mathbb{R}}^2(\mathbb{N})$.

Finally, letting $m \to \infty$ in equation (3) we obtain $\sum_{j=1}^{\infty}(x_{nj} - x_{0j})^2 \leq \varepsilon^2$ when $n \geq N$, and this says that $\|\vec{x}_0 - \vec{x}_n\|_2 < \varepsilon$ when $n \geq N$.

Exercises 4

*1. (a) Let $(V, \langle,\rangle)$ be an inner product space and let S be an orthonormal subset of V. Show that S is a maximal orthonormal set if, and only if, the only vector in V that is orthogonal to every element of S is the zero vector.

(b) For each $n \in \mathbb{N}$ let $\vec{e}_n = (0, \ldots, 0, 1, 0, \ldots)$ where the one is in the nth place. Clearly, $\{\vec{e}_n\}_{n=1}^{\infty} \subseteq \ell_{\mathbb{R}}^2(\mathbb{N})$ and also $\{\vec{e}_n\} \subseteq \ell_c^2(\mathbb{N})$ (Section 1, examples (ii) and (iii)). Show that this set is a maximal orthonormal sequence in $\ell_{\mathbb{R}}^2(\mathbb{N})$ and also in $\ell_c^2(\mathbb{N})$.

(c) Recall the inner product on $C(T)$ (Exercises 3, problem 4). Show that the orthonormal sequence $\left\{\dfrac{e^{int}}{\sqrt{2\pi}} \,\Big|\, n \in \mathbb{Z}\right\}$ is maximal. Hint: See Chapter One, Section 7.

2. Let $(H_1, \langle,\rangle_1)$ and $(H_2, \langle,\rangle_2)$ be two Hilbert spaces over the same field. Suppose that $\{\vec{v}_j\}_{j=1}^{\infty} \subseteq H_1$ and $\{\vec{w}_j\}_{j=1}^{\infty} \subseteq H_2$ are maximal orthonormal sequences (let us stress that both of these are infinite sequences). For each j let $T(\vec{v}_j) = \vec{w}_j$ and, given $\vec{u} = \sum_{j=1}^{\infty} \alpha_j \vec{v}_j$ in H_1, define $T(\vec{u})$ to be $\sum_{j=1}^{\infty} \alpha_j \vec{w}_j$.

   (a) Show that T is well defined.
   (b) Show that T is a linear one-to-one mapping from H_1 onto H_2.
   (c) For any vectors $\vec{u}, \vec{v}$ in H_1 show that $\langle \vec{u}, \vec{v} \rangle_1 = \langle T(\vec{u}), T(\vec{v}) \rangle_2$; in particular, $\|\vec{u}\|_1 = \|T(\vec{u})\|_2$ for every $\vec{u} \epsilon H_1$.

5. The Completion

The Lebesgue integral had a profound effect on the study of trigonometric series. It led, among many other things, to the creation of the Lebesgue classes; also known as the L^p spaces (read "L.P."). Here $1 \leq p < \infty$, and for each such p, L^p is a Banach space whose elements are equivalence classes (Chapter Zero, Exercises 2, problem 1) of functions defined on, let us say, $[-\pi, \pi]$. For $f \epsilon L^p[-\pi, \pi]$ the L^p-norm of f is defined as follows:

$$\|f\|_p = \left(\int_{-\pi}^{\pi} |f(t)|^p \, dt \right)^{\frac{1}{p}}$$

In this context, equal functions have the same integral, so this is well defined. It is customary to "abuse language" and to refer to the elements of L^p as L^p-functions. The connection to trigonometric series is this:

Theorem 1. For each fixed p, with $1 \leq p < \infty$, and each $f \epsilon L^p[-\pi, \pi]$ the Cesàro means of the Fourier series of f converge to this function for the L^p-norm.

When $p = 2$, and only then, $\|\cdot\|_2$ is the norm associated with an inner product. Thus $L^2[-\pi, \pi]$ is a Hilbert space and here we have a stronger result:

Theorem 2. The Fourier series of any function in $L^2[-\pi, \pi]$ converges to that function for the L^2-norm.

We shall not prove Theorem 1 (see the Notes for a reference). Theorem 2 follows immediately once we show that $\left\{ \frac{e^{int}}{\sqrt{2\pi}} \right\}_{-\infty}^{\infty}$ is an orthonormal basis for $L^2[-\pi, \pi]$ (Section 4, Theorem 2). We shall do this very soon.

Since we are not assuming a knowledge of the Lebesgue integral, so we must define the L^p-spaces in a different way.

Definition 1. Let $(E, \|\cdot\|)$ be a normed space and let D be a subset of E. We shall say that D is a dense subset of E or that D is dense in E if for any $\vec{x} \epsilon E$ we can find a sequence $\{\vec{y}_n\} \subseteq D$ such that $\lim_{n\to\infty} \vec{y}_n = \vec{x}$. If $(V, \langle,\rangle)$ is an inner product space and $D \subseteq V$, we shall say that D is dense in V if it is dense in $(V, \|\cdot\|)$ where $\|\cdot\|$ is the norm associated with $\langle,\rangle$.

We recall that the set of trigonometric polynomials (Chapter One, Section 7, just before Corollary 2) is dense in $(C(T), \|\cdot\|_\infty)$. Theorem 1, quoted above, shows that $C(T)$ is dense in $(L^p[-\pi, \pi], \|\cdot\|_p)$ for each p. We shall use this fact to define the L^p-spaces.

Theorem 3. For any normed space $(E, \|\cdot\|)$ there is a Banach space $(B_1, \|\cdot\|_1)$ and a linear map $T : E \to B_1$ such that:

(a) For every $\vec{x} \epsilon E, \|\vec{x}\| = \|T(\vec{x})\|_1$;

(b) The set (it is actually a linear subspace) $T(E) = \{T(\vec{x}) \mid \vec{x} \epsilon E\}$ is dense in $(B_1, \|\cdot\|_1)$.

Furthermore, if $(B_2, \|\cdot\|_2)$ is another Banach space that has properties (a) and (b), then there is a linear map $T' : B_1 \to B_2$ that is one-to-one, onto, and norm preserving (i.e., $\|T'(\vec{x})\|_2 = \|\vec{x}\|_1$ for every $\vec{x} \epsilon B_1$).

A proof of this theorem can be found in Appendix B. We call $(B_1, \|\cdot\|_1)$ the completion of $(E, \|\cdot\|)$ and often say, identifying E with $T(E)$, that E is dense in its completion.

On the space $C(T)$ we can define a function $\|\cdot\|_1$ by setting

$$\|f\|_1 = \int_{-\pi}^{\pi} |f(t)| \, dt$$

for every $f \epsilon C(T)$. It is easy to see that this gives us a norm on $C(T)$. The completion of $(C(T), \|\cdot\|_1)$ is defined to be the Banach space $L^1[-\pi, \pi]$. When $p > 1$, and $p \neq 2$, it is *not* so easy to show that $\|\cdot\|_p$, defined to be

$$\left(\int_{-\pi}^{\pi} |f(t)|^p \, dt\right)^{\frac{1}{p}}$$

for each $f \epsilon C(T)$, is a norm. It is, however, and the completion of $(C(T,) \|\cdot\|_p)$ is defined to be $L^p[-\pi, \pi]$. Because we do not need these spaces, we do not carry this discussion any further.

For the special case $p = 2$ we have a stronger form of Theorem 3:

Theorem 4. Given an inner product space $(V \langle,\rangle)$ there is a Hilbert space $(H_1, \langle\rangle_1)$ and a linear map $T : V \to H_1$ such that

(a) For all $\vec{u}, \vec{v}$ in V, $\langle T\vec{u}, T\vec{v}\rangle_1 = \langle \vec{u}, \vec{v}\rangle$;

(b) The set (it is actually a linear subspace) $T(V) = \{T(\vec{v}) \mid \vec{v} \epsilon V\}$ is dense in $(H_1, \langle,\rangle_1)$.

Furthermore, if $(H_2, \langle,\rangle_2)$ is another Hilbert space that has properties (a) and (b), then there is a linear map $T' : H_1 \to H_2$ that is one-to-one, onto, and for which $\langle T'\vec{u}, T'\vec{v}\rangle_2 = \langle \vec{u}, \vec{v}\rangle_1$ for all $\vec{u}, \vec{v}$ in H_1.

A proof of this theorem can be found in Appendix B. We call $(H_1, \langle,\rangle_1)$ the completion of $(V, \langle,\rangle)$ and, as in the case of a normed space, we say V is dense in its completion.

We have defined for any $f, g \epsilon C(T)$, $\langle f, g\rangle = \int_{-\pi}^{\pi} f(t) \overline{g(t)} dt$, and we have seen that this is an inner product $C(T)$. The completion of $(C(T), \langle,\rangle)$ is denoted by $L^2(T)$ (read "L two of T").

The inner product and norm of $L^2(T)$, when restricted to elements of $C(T)$, is just the inner product and norm as defined above. Thus the sequence $\left\{\dfrac{e^{int}}{\sqrt{2\pi}}\right\}_{-\infty}^{\infty}$ is an orthonormal sequence in $L^2(T)$. We shall show that it is an orthonormal basis for this Hilbert space. First we need:

Lemma 1. An orthonormal sequence $\{\vec{v}_j\}$ in a Hilbert space $(H, \langle,\rangle)$ is an orthonormal basis for H if, and only if, the linear span of $\{\vec{v}_j\}$ (Appendix A) is dense in H.

Proof. Recall that the linear span of $\{\vec{v}_j\}$, denoted by $lin\{\vec{v}_j\}$, is the set of all (finite) linear combinations of elements of $\{\vec{v}_j\}$.

Suppose first that $\{\vec{v}_j\}$ is an orthonormal basis for H, and let $\vec{v} \epsilon H$ be given. Then by part (b) of Theorem 2 (Section 4) we have

$$\vec{v} = \sum_{j=1}^{\infty} \langle \vec{v}, \vec{v}_j \rangle > \vec{v}_j$$

Where convergence is for the norm of H. But then the sequence

$$\left\{\sum_{j=1}^{n} \langle \vec{v}, \vec{v}_j \rangle \vec{v}_j\right\}_{n=1}^{\infty}$$

which is clearly contained in $lin\{\vec{v}_j\}$, converges to $\vec{v}$ showing that the linear span is dense in H.

Now suppose that $lin\{\vec{v}_j\}$ is dense in H. To show that $\{\vec{v}_j\}$ is an orthonormal basis for H, we need only show that it is maximal (Section 4, Theorem 2).

So suppose $\vec{v} \epsilon H$ and $\langle \vec{v}, \vec{v}_j \rangle = 0$ for every j. We must show that $\vec{v}$ is the zero vector. Let $\varepsilon > 0$ be given. By our assumption there is a $\vec{w} \epsilon\, lin\{\vec{v}_j\}$ such that $\|\vec{v} - \vec{w}\| < \varepsilon$. Thus, by Theorem 2 of Section 3, we have

$$\|\vec{v}\|^2 = |\langle \vec{v}, \vec{v} \rangle| = |\langle \vec{v} - \vec{w}, \vec{v} \rangle| \le \|\vec{v} - \vec{w}\|\|\vec{v}\| < \varepsilon \|\vec{v}\|$$

because $\vec{w} \epsilon\, lin\{\vec{v}_j\}$ and $\langle \vec{v}, \vec{v}_j \rangle = 0$ for all d so $\langle \vec{v}, \vec{w} \rangle = 0$. Now if $\vec{v}$ is not the zero vector, then $\|\vec{v}\| < \varepsilon$ for any $\varepsilon > 0$, which is clearly a contradiction.

Theorem 5. The set $\left\{\dfrac{e^{int}}{\sqrt{2\pi}}\right\}_{-\infty}^{\infty}$ is an orthonormal basis for the Hilbert space $L^2(T)$.

Proof. According to Lemma 1 we need only show that the linear span of the given sequence is dense in $L^2(T)$. Let $f \epsilon L^2(T)$ and $\varepsilon > 0$ be given and use the fact that $C(T)$ is dense in this space to find $g \epsilon C(T)$ such that $\|f - g\|_2 < \frac{\varepsilon}{2}$. Now the linear span of $\left\{\dfrac{e^{int}}{\sqrt{2\pi}}\right\}_{-\infty}^{\infty}$ is dense in $(C(T), \|\cdot\|_\infty)$ (Chapter One, Corollary 2, Section 7). Thus there is an element τ of this linear span such that $\|g - \tau\|_\infty < \dfrac{\varepsilon}{2\sqrt{2\pi}}$. But then $\|g - \tau\|_2^2 = \int_{-\pi}^{\pi} |g(t) - \tau(t)|^2\, dt < \dfrac{\varepsilon^2}{4(2\pi)}(2\pi) = \dfrac{\varepsilon^2}{4}$.

Combining our inequalities we have $\|f - \tau\|_2 \le \|f - g\|_2 + \|g - \tau\|_2 < \frac{\varepsilon}{2} + \frac{\varepsilon}{2} = \varepsilon$ and we are done.

Corollary 1. For any $f \epsilon L^2(T)$ the series $\displaystyle\sum_{-\infty}^{\infty} \left\langle f, \dfrac{e^{int}}{\sqrt{2\pi}} \right\rangle \dfrac{e^{int}}{\sqrt{2\pi}}$ converges to f for the L^2-norm. In particular for any $f \epsilon (C(T))$, the series

$$\sum_{-\infty}^{\infty} \left\langle f, \dfrac{e^{int}}{\sqrt{2\pi}} \right\rangle \dfrac{e^{int}}{\sqrt{2\pi}} = \sum_{-\infty}^{\infty} \left(\dfrac{1}{2\pi} \int_{-\pi}^{\pi} f(t) e^{-int}\, dt \right) e^{int}$$

converges to f for the L^2-norm. For any such f we also have

$$2\pi \sum_{-\infty}^{\infty} |c_n(f)|^2 = \int_{-\pi}^{\pi} |f(t)|^2\, dt = \|f\|_2^2$$

Exercises 5

*1. There is no inner product on $L^1[-\pi,\pi]$. Still, $\{e^{int}\}_{-\infty}^{\infty}$ is contained in this space. Show that the linear span of this sequence is dense in $(L^1[-\pi,\pi], \|\cdot\|_1)$.

*2. When $f \epsilon C(T)$ the numbers $\alpha_n = \left\langle f, \dfrac{e^{int}}{\sqrt{2\pi}} \right\rangle$ satisfy $\sum_{-\infty}^{\infty} |\alpha_n|^2 < \infty$.

   Given a sequence of numbers $\{\beta_n\}_{-\infty}^{\infty}$ such that $\sum_{-\infty}^{\infty} |\beta_n|^2 < \infty$ we may ask if there is an $f \epsilon C(T)$ such that $\left\langle f, \dfrac{e^{int}}{\sqrt{2\pi}} \right\rangle = \beta_n$ for any n. The answer is "no." Show, however, that there is an element $g \epsilon L^2(T)$ such that $\left\langle g, \dfrac{e^{int}}{\sqrt{2\pi}} \right\rangle = \beta_n$ for all n.

*3. Let $C(\mathbb{R})$ be the set of all complex-valued functions that are defined and continuous on $\mathbb{R}$.

   (a) Show that $C(\mathbb{R})$ is a vector space over $\mathbb{C}$ (see Section 1 for a discussion of the algebraic operations on functions).

   (b) An element $f \epsilon C(\mathbb{R})$ is said to have compact support if there are real numbers a, b such that $\{x \epsilon \mathbb{R} \mid f(x) \neq 0\} \subseteq [a,b]$.

   Let $C_0(\mathbb{R})$ be the set of all functions in $C(\mathbb{R})$ that have compact support. Show that $C_0(\mathbb{R})$ is a vector space over $\mathbb{C}$.

   (c) For each $f \epsilon C_0(\mathbb{R})$ let $\|f\|_1 = \int_{-\infty}^{\infty} |f(x)|\,dx$. Show that $\|\cdot\|_1$ is a norm on $C_0(\mathbb{R})$. The completion of $(C_0(\mathbb{R}), \|\cdot\|_1)$ is a Banach space that we shall denote by $L^1(\mathbb{R})$ (read "L one of $\mathbb{R}$").

4. For each f, g in $C_0(\mathbb{R})$ define $\langle f, g \rangle$ to be $\int_{-\infty}^{\infty} f(x)\overline{g(x)}\,dx$. Show that $\langle,\rangle$ is an inner product on $C_0(\mathbb{R})$. The completion of $(C_0(\mathbb{R}), \langle,\rangle)$ is a Hilbert space that we shall denote by $L^2(\mathbb{R})$.

6. Wavelets

One of the major applications of Fourier series is to the analysis of real-world data. We have seen that $\left\{ \dfrac{e^{int}}{\sqrt{2\pi}} \mid n \in Z \right\}$ is an orthonormal basis for $L^2[0, 2\pi]$ (Section 5, Theorem 5) and so any function in this space can be approximated, as closely as we wish in the L^2 norm, by a finite linear combination (what physicists call a "superposition") of sines and cosines. There are many instances, however, when such sums fail to reflect some of the subtleties in the data. A better tool, in these instances, are the wavelet bases. This subject can be traced to some research in pure mathematics; however, many important results came from the investigations of geophysicists and others interested in signal processing. There is now an extensive literature on wavelets and their many important applications, and we can only give the reader a very small sample of the mathematical ideas underlying the subject.

A string of measurements, one-dimensional data, or a signal can be thought of as a function of time, $f(t)$. The energy in the signal is given by $\int_{-\infty}^{\infty} |f(t)|^2 \, dt$ and because this is finite, $f \in L^2(\mathbb{R})$ (Exercises 5, problem 4). In many cases of interest f has compact support (Exercises 5, problem 3b) and many sharp spikes or sharp discontinuities. To represent such functions in a series, we seek a basis for $L^2(\mathbb{R})$, but we want our basis to represent our function at different scales (resolutions). In this way both gross or large-scale features and small subtle features of our data are captured in the representation.

Definition 1. A wavelet is a function $\psi(t) \in L^2(\mathbb{R})$ such that $\left\{ 2^{\frac{j}{2}} \psi \left(2^j t - k \right) \mid j, k \in Z \right\}$ is an orthonormal basis for $L^2(\mathbb{R})$.

Note that we have here both translations (the role of k) and dilations of the function ψ, the so-called mother wavelet. At this point it is not at all clear that any wavelets exist. There are many, however, and, in applications, a judicious choice of ψ yields a powerful new way to analyze data.

Wavelets can be constructed from a multiresolution analysis. This is, by definition, a sequence of closed linear subspaces $\{V_j\}$ of $L^2(\mathbb{R})$ such that

(i) $\subset V_{-1} \subset V_0 \subset V_1 \subset ...$;

(ii) $\cup \{V_j \mid j \in Z\}$ is dense (Section 5, Definition 1) in $L^2(\mathbb{R})$;

(iii) $\cap \{V_j \mid j \in Z\} = \{0\}$;

(iv) $f(t) \in V_j$ if, and only if, $f(2^{-j} t) \in V_0$;

(v) There is a function $\varphi \in V_0$, called the scaling function, such that $\{\varphi(t-k) \mid k \in Z\}$ is an orthonormal basis for V_0.

In applications, a multiresolution analysis is a mathematical framework for decomposing a signal into components of different scales. The spaces V_n correspond to these scales.

If any V_n is given then by (iv), any other V_m can be found because $V_m = \{f(2^{m-n} t) \mid f \in V_n\}$. Hence, a multiresolution analysis is completely determined by the scaling function φ (sometimes called the "father wavelet"). Because the translates of φ give us an orthonormal basis for V_0, the translates of $\varphi(2^n t)$ gives us an orthonormal basis for each V_n. Unfortunately, however, the union of these bases is not, in general, a basis for $L^2(\mathbb{R})$. To obtain an orthonormal basis of wavelets, we construct a new sequence of subspaces as follows. For each fixed $n \in Z$ let $W_n = \{\vec{w} \in V_{n+1} \mid \langle \vec{w}, \vec{v} \rangle = 0 \text{ for all } \vec{v} \in V_n\}$; W_n is called the orthogonal complement of V_n in V_{n+1}. Every vector in V_{n+1} can be written, in just one way, as a sum of one vector from V_n and one from W_n (proved below). To indicate this, we write $V_{n+1} = V_n \oplus W_n$.

We can repeat this process. $V_n = V_{n-1} \oplus W_{n-1}$ and so $V_{n+1} = V_{n-1} \oplus W_{n-1} \oplus W_n$. Continuing in this way we find that every vector in $L^2(\mathbb{R})$ can be written as an infinite sum of vectors one from each of the spaces W_n. In symbols $L^2(\mathbb{R}) = \bigoplus_{n=-\infty}^{\infty} W_n$. These spaces are mutually orthogonal and $f(t) \in W_n$ if, and only if, $f(2^{-n}t) \in W_0$. We see from this last fact that once we find a function whose translates give us an orthonormal basis for W_0, we can obtain from that function an orthonormal basis for every W_n, and because of the mutual orthogonality of the W_n the union of these bases is an orthonormal basis for $L^2(\mathbb{R})$.

The problem of finding a wavelet can be now outlined as follows. Given a multiresolution analysis $\{V_n\}$, and recall this means we are given a scaling function φ whose translates constitute an orthonormal basis for V_0, construct the sequence $\{W_n\}$ and find a function ψ whose translates constitute an orthonormal basis for W_0. The function ψ is a wavelet. It turns out that such a ψ exists for every multiresolution analysis, and the proof of this fact actually gives a method of constructing ψ. We have:

Theorem 1. Let $\{V_n\}$ be a multiresolution analysis with the scaling function φ. Then the function $\psi(t) = \sum_{-\infty}^{\infty} (-1)^k \overline{h}_{1-k} \varphi(2t-k)$, where $h_k = 2 \int_{-\infty}^{\infty} \varphi(t) \overline{\varphi(2t-k)} dt$ is a mother wavelet; i.e., $\{2^{\frac{n}{2}} \psi(2^n t - m) \mid m, n \in Z\}$ is an orthonormal basis for $L^2(\mathbb{R})$.

Perhaps the simplest example of the process discussed above is found by using the Haar scaling function $H(t)$, where $H(t) = 1$ for $0 \leq t < 1$ and $H(t) = 0$ for all other t. It is easy to see that the integer translates of $H(t)$ give us an orthonormal set in $L^2(\mathbb{R})$. These translates generate the space V_0.

The function $H(t)$ has its support in $[0,1]$, whereas the function $H(2t-n)$ has its support in $\left[\frac{n}{2}, \frac{n+1}{2}\right]$. These overlap when $n=0$ in the set $\left[0, \frac{1}{2}\right]$ and when $n=1$ in the set $\left[\frac{1}{2}, 1\right]$. Thus for the Haar function $h_0 = h_1 = 1$ and $\psi(x) = \varphi(2x) - \varphi(2x-1)$.

The function $\psi(x)$ is called the Haar wavelet. As we have already said, there are many others and not all of them have compact support. For more about this fascinating subject and for a proof of Theorem 1, we refer the reader to the references listed in Notes.

We end this section with a proof of a property of Hilbert spaces that played a key role in the construction above.

Lemma 1. Let $(H, <,>)$ be a Hilbert space, let M be a closed linear subspace of H, and let $\vec{y} \in H \setminus M$. Then there is a unique vector $\vec{m}_0 \in M$ such that $\left\|\vec{y} - \vec{m}_0\right\| = \inf\left\{\left\|\vec{y} - \vec{m}\right\| \mid \vec{m} \in M\right\}$. Furthermore, $\left\langle \vec{y} - \vec{m}_0, \vec{m} \right\rangle = 0$ for all $\vec{m} \in M$.

Proof. Let $d = \inf\left\{\left\|\vec{y} - \vec{m}\right\| \mid \vec{m} \in M\right\}$ and choose a sequence $\left\{\vec{m}_n\right\}_{n=1}^{\infty} \subseteq M$ such that $d = \lim_{n \to \infty} \left\|\vec{y} - \vec{m}_n\right\|$. Then, by Exercises 3, problem 1b, we may write

$$\left\|\left(\vec{y} - \vec{m}_k\right) + \left(\vec{y} - \vec{m}_l\right)\right\|^2 + \left\|\left(\vec{y} - \vec{m}_k\right) - \left(\vec{y} - \vec{m}_l\right)\right\|^2$$

$= 2\left\|\vec{y} - \vec{m}_k\right\|^2 + 2\left\|\vec{y} - \vec{m}_l\right\|^2$. We may write this as

$$\left\|\vec{m}_l - \vec{m}_k\right\| = 2\left\|\vec{y} - \vec{m}_k\right\|^2 + 2\left\|\vec{y} - \vec{m}_l\right\|^2 - \left\|2\vec{y} - \left(\vec{m}_l + \vec{m}_k\right)\right\|^2.$$

The very last term in this expression can be written as $4\left\|\vec{y} - \dfrac{\vec{m}_l + \vec{m}_k}{2}\right\|^2$.

Now M is a linear subspace and $\vec{m}_l, \vec{m}_k$ are in M, so $\dfrac{\vec{m}_l + \vec{m}_k}{2}$ is in M. It follows that $4\left\|\vec{y} - \dfrac{\vec{m}_l + \vec{m}_k}{2}\right\|^2 \geq 4d^2$ and so

$$\left\|\vec{m}_l - \vec{m}_k\right\|^2 \leq 2\left\|\vec{y} - \vec{m}_k\right\|^2 + 2\left\|\vec{y} - \vec{m}_l\right\|^2 - 4d^2.$$

Recalling the way $\left\{\vec{m}_n\right\}$ was defined we see that our last inequality implies that $\left\{\vec{m}_n\right\}$ is a Cauchy sequence (Section 2, Definition 3). Because H is a Hilbert space this sequence converges, because $\left\{\vec{m}_n\right\} \subseteq M$ and M is closed, its limit, say $\vec{m}_0$, is in M (Section 2, Lemma 1). Thus $\left\|\vec{y} - \vec{m}_0\right\| = \lim \left\|\vec{y} - \vec{m}_n\right\| = d$.

Let us now show that $\vec{m}_0$ is unique. Suppose we had two vectors $\vec{m}_1, \vec{m}_2$, in M such that $\left\|\vec{y} - \vec{m}_1\right\| = d = \left\|\vec{y} - \vec{m}_2\right\|$. Then, as above

$$\left\|\left(\vec{y} - \vec{m}_1\right) + \left(\vec{y} - \vec{m}_2\right)\right\|^2 + \left\|\left(\vec{y} - \vec{m}_1\right) - \left(\vec{y} - \vec{m}_2\right)\right\|^2$$

$= 2\left\|\vec{y} - \vec{m}_1\right\|^2 + 2\left\|\vec{y} - \vec{m}_2\right\|^2$. This may be written as

$$\left\|\vec{m}_2 - \vec{m}_1\right\|^2 = 4d^2 - 4\left\|\vec{y} - \frac{\vec{m}_1 + \vec{m}_2}{2}\right\|,$$ but as we discussed in the first part of the proof $4\left\|\vec{y} - \frac{\vec{m}_1 + \vec{m}_2}{2}\right\|^2 \geq 4d^2$ and so $\left\|\vec{m}_2 - \vec{m}_1\right\|^2 \leq 0$, giving us $\vec{m}_2 = \vec{m}_1$.

Finally, we must show that $\langle \vec{y} - \vec{m}_0, \vec{m} \rangle = 0$ for all $\vec{m} \in M$. Let us set $\vec{y}_0 = \vec{y} - \vec{m}_0$. Note that for any $\vec{m} \in M$ and any scalar δ, $\vec{m}_0 + \delta \vec{m}$ is in M, and so $d^2 \leq \left\|\vec{y} - (\vec{m}_0 + \delta \vec{m})\right\|^2 = \left\|\vec{y}_0 - \delta \vec{m}\right\|^2 = \langle \vec{y}_0 - \delta \vec{m}, \vec{y}_0 - \delta \vec{m} \rangle$

$= \left\|\vec{y}_0\right\|^2 - \delta \langle \vec{m}, \vec{y}_0 \rangle - \overline{\delta} \langle \vec{y}_0, \vec{m} \rangle + |\delta|^2 \left\|\vec{m}\right\|^2.$

Because $\left\|\vec{y}_0\right\|^2 = \left\|\vec{y} - \vec{m}_0\right\|^2 = d^2$ this becomes

$(*)\ 0 \leq -\delta \langle \vec{m}, \vec{y}_0 \rangle - \overline{\delta} \langle \vec{y}_0, \vec{m} \rangle + |\delta|^2 \left\|\vec{m}\right\|^2.$

Now suppose $\langle \vec{y}_0, \vec{m} \rangle \neq 0$ for some $\vec{m} \in M$. Then $\vec{m}$ could not be zero (Exercises 3, problem 1a) and so we may set $\delta = \dfrac{\langle \vec{y}_0, \vec{m} \rangle}{\left\|\vec{m}\right\|^2}$.

Putting this into $(*)$ we obtain

$0 \leq -2 \dfrac{\left|\langle \vec{y}_0, \vec{m} \rangle\right|^2}{\left\|\vec{m}\right\|^2} + \dfrac{\left|\langle \vec{y}_0, \vec{m} \rangle\right|^2}{\left\|\vec{m}\right\|^2} = -\dfrac{\left|\langle \vec{y}_0, \vec{m} \rangle\right|^2}{\left\|\vec{m}\right\|^2}$, which is a contradiction.

Theorem 2. Let $(H, \langle , \rangle)$ be a Hilbert space and let M be a closed linear subspace of H. Then $M^\perp = \left\{\vec{v} \in H \mid \langle \vec{v}, \vec{m} \rangle = 0 \text{ for all } \vec{m}\right\}$ is a closed linear subspace of H. Furthermore, (a) $M \cap M^\perp = \{\vec{0}\}$, and (b) each $\vec{v} \in H$ can be written in one, and only one, way as a sum of a vector in M and a vector in $M^\perp$.

Proof. The proof that $M^\perp$ is a closed linear subspace we leave to the reader (Exercises 6, problem 1). If $\vec{v} \in M \cap M^\perp$, then $\langle \vec{v}, \vec{v} \rangle = 0$, which gives $\vec{v} = \vec{0}$ (Section 3, Definition 1c).

For any $\vec{v} \in H \setminus M$, there is a unique $\vec{m}_0 M$ such that $\left\|\vec{v} - \vec{m}_0\right\| = \inf\left\{\left\|\vec{v} - \vec{m}\right\| \mid \vec{m} \in M\right\}$ and, as we showed in Lemma 1, $\langle \vec{v} - \vec{m}_0, \vec{m} \rangle = 0$ for all $\vec{m} \in M$. Thus $\vec{v} - \vec{m}_0$ is in $M^\perp$, i.e., $\vec{v} - \vec{m}_0 = \vec{p}$ for some vector $\vec{p} \in M^\perp$. It follows then that $\vec{v} = \vec{m_0} + \vec{p}$, where $\vec{m_0} \in M, \vec{p} \in M^\perp$.

Suppose now that $\vec{v} = \vec{m_0}' + \vec{p}'$ where $\vec{m_0}' \in M, \vec{p}' \in M^\perp$. Then $\vec{m_0} + \vec{p} = \vec{m_0}' + \vec{p}'$ so $\left(\vec{m_0} - \vec{m_0}'\right) = \left(\vec{p} - \vec{p}'\right)$. Call this vector $\vec{h}$. Clearly $\vec{h} \in M$, because $\vec{h} = \vec{m_0} - \vec{m_0}'$, and $\vec{h} \in M^\perp$, because $\vec{h} = \vec{p}' - \vec{p}$. Thus, by (a) $\vec{h} = \vec{0}$ and we have $\vec{m_0} = \vec{m_0}', \vec{p} = \vec{p}'$.

Exercises 6

1. Let $(H, \langle, \rangle)$ be a Hilbert space and let M be a closed linear subspace of H.

   (a) Show that $M^\perp = \{w \in H \mid \langle w, v \rangle = 0 \text{ for all } V \in m\}$ is a closed linear subspace of H.

   (b) Define $M^{\perp\perp}$ to be $\left(M^\perp\right)^\perp$ and show that $M^{\perp\perp} = M$.

Notes

1. A discussion of the foundations of quantum mechanics can be found in *Introduction to Hilbert Spaces with Applications,* by Lokenath Debnath and Piotr Mikusinski (Academic Press, 2nd edition, New York, 1999). See Chapter 7.

2. A proof of Theorem 1 (Section 5) can be found in *Functional Analysis,* by Carl L. DeVito (Academic Press, New York, 1978), pp. 92–93.

3. Wavelets are discussed in the book listed in note 1 above. See Chapter 8, and for a proof of Theorem 1 (Section 6) see page 441 (their notation is slightly different from ours.) Wavelets are also treated in *Wavelets and Operators,* by Yves Meyer (Cambridge University Press, Cambridge, 1992), in *A Mathematical Introduction to Wavelets,* by P. Wojtaszczyk (Cambridge University Press, Cambridge, 1997), and in *Wavelets—Algorithms and Applications,* by Yves Meyer, translated and revised by Robert D. Ryan (Society for Industrial and Applied Mathematics, Philadelphia, 1993). This last reference contains an interesting discussion of the functions of Riemann and Weierstrass (see the discussion at the end of Section 4, Chapter 1) and their connections to wavelets. See pages 115–118.

Chapter Four
The Fourier Transform

In the last three chapters we have been concerned with the Fourier series and the problem of recapturing a function from its Fourier series. Here we shall change our point of view and investigate the properties of a certain mapping between pairs of "function spaces." By the term "function space" we mean a collection of functions that is a vector space over $\mathbb{C}$ when addition and scalar multiplication are defined in the "usual" way. The mapping, called the Fourier transform, is the object of interest here.

There are a number of natural questions that arise at this point. Why change the emphasis in this way? What, if any, is the advantage in doing so? At this early stage these questions are hard to answer. Suffice it to say that the Fourier transform plays a central role in many areas of engineering, of physics, of chemistry, and even, in recent times, of astronomy. We discuss some of these applications later. So the effort involved in understanding this mapping is well justified. Moreover, the associated mathematics is fascinating and very beautiful.

1. The Fourier Transform on Z

We recall the space $\ell^1(Z) = \{a : Z \to \mathbb{C} \mid \Sigma |a(n)| < \infty\}$ defined in Chapter Three, Section 1, example iv, and the space $C(T)$ of continuous functions on the circle. There is a natural mapping between these spaces: For each $a \in \ell^1(Z)$ we set $\Im(a) = \sum_{-\infty}^{\infty} a(n)e^{in\theta}$ and note that because $\sum |a(n)| < \infty$, the series converges uniformly over T to an element of $C(T)$.

The map $\Im$ is called the Fourier transform on Z. Its properties are easily proved, and so our results here provide us with a model for the discussion of this mapping in more complicated settings; i.e., we shall seek analogues of the theorems proved here for the Fourier transform on $\mathbb{R}$ and on Z_n (defined later).

Lemma 1. The map $\Im : \ell^1(Z) \to C(T)$ is linear and has an inverse.

Proof. For any a, b in $\ell^1(Z)$ we have $(a+b)(n) = a(n) + b(n)$ for all $n \in Z$. Thus

$$\Im(a+b) = \Sigma[(a+b)(n)]e^{in\theta} = \Sigma[a(n) + b(n)]e^{in\theta}$$
$$= \Sigma[a(n)e^{in\theta} + b(n)e^{in\theta}]$$
$$= \Sigma a(n)e^{in\theta} + \Sigma b(n)e^{in\theta} = \Im(a) + \Im(b)$$

It is easy to see that given $\lambda \in \mathbb{C}$, $\Im(\lambda a) = \lambda \Im(a)$ for every $a \in \ell^1(Z)$ and so $\Im$ is a linear mapping.

We have already noted that for any $a \in \ell^1(Z)$ the series $\Sigma a(n)e^{in\theta}$ converges uniformly over T. If we denote this sum by $\hat{a}(\theta)$, then (Chapter One, Section 5) it follows that

$$a(n) = \frac{1}{2\pi} \int_{-\pi}^{\pi} \hat{a}(\theta) e^{-in\theta} d\theta, \quad \text{for every } n \in Z$$

Thus $\Im$ is one-to-one and, in fact, $\Im^{-1}$ exists and is given by

$$\Im^{-1}[\hat{a}(\theta)] = \{\frac{1}{2\pi} \int_{-\pi}^{\pi} \hat{a}(\theta) e^{-in\theta} d\theta\}_{n=-\infty}^{\infty}$$

It is convenient to denote $\Im(a)$ by $\hat{a}(\theta)$, or just $\hat{a}$, and we shall often use this notation below.

Recall that $\ell^1(Z)$ and $C(T)$ both have norms (Chapter Three, Section 2, Definition 1). For each $a \in \ell^1(T)$, $\|a\|_1 = \sum |a(n)|$, and for each $f \in C(T)$, $\|f\|_\infty = \max_{\theta \in T} |f(\theta)|$.

Lemma 2. The map $\Im : \ell^1(Z) \to C(T)$ is continuous and, in fact, is norm decreasing; i.e., $\|\hat{a}\|_\infty \leq \|a\|_1$ for all a.

Proof.

$$\|\Im(a)\|_\infty = \|\hat{a}\|_\infty = \max_{\theta \in T} |\sum a(n)e^{in\theta}|$$
$$\leq \sum |a(n)| \|e^{in\theta}\|_\infty \leq \sum |a(n)| = \|a\|_1$$

The continuity of $\Im$ now follows from Theorem 1 of Chapter Three, Section 2.

There is a very useful operation on $\ell^1(Z)$ called "convolution." It is a kind of multiplication.

Definition 1. For any a, b in $\ell^1(Z)$ and any n in Z, we define $(a*b)(n)$ to be $\sum_{m=-\infty}^{\infty} a(n-m)b(m)$. This sequence $\{(a*b)(n)\}_{n=-\infty}^{\infty}$ is called the convolution of a and b, or a convolved with b, and is denoted by $a*b$.

We must show that $a * b$ is in $\ell^1(Z)$ whenever a and b are both in this space. The next lemma proves that and gives us a bound for the norm of the convolution.

Lemma 3. For any a, b in $\ell^1(Z)$ the convolution, $a*b$, is in $\ell^1(Z)$ and $\|a*b\|_1 \leq \|a\|_1 \cdot \|b\|_1$.

Proof. Because $\|a\|_1 = \sum_{-\infty}^{\infty} |a(n)|$ we have, for any fixed m, $\|a\|_1 = \sum |a(n-m)|$. Now

$$\|a * b\|_1 = \sum_{-\infty}^{\infty} |(a*b)(n)| = \sum_n |\sum_m a(n-m)b(m)|$$

$$\leq \sum_n \sum_m |a(n-m)| \, |b(m)| = \sum_m |b(m)| (\sum_n |a(n-m)|)$$

$$= \|a\|_1 \sum_m |b(m)| = \|a\|_1 \|b\|_1$$

The function $e(n) = 1$ when $n = 0$, $e(n) = 0$ when $n \neq 0$ is clearly in $\ell^1(Z)$. Note that $a * e = e * a = a$ for all a and, further, $\widehat{e}(\theta) = 1$ for all $\theta \in T$.

Two functions f, g in $C(T)$ can be multiplied. We simply set $(f \cdot g)(\theta) = f(\theta)g(\theta)$ for all $\theta \in T$ and observe that $f \cdot g \in C(T)$. The Fourier transform, besides preserving sums and scalar multiples (Lemma 1), also preserves "multiplication." More precisely:

Lemma 4. For any a, b in $\ell^1(Z)$, $\Im(a * b) = \Im(a) \cdot \Im(b)$. In another notation: $\widehat{a * b} = \widehat{a} \cdot \widehat{b}$.

Proof. In the expression

$$\widehat{a * b}(\theta) = \sum_n \left\{ \sum_m a(n-m)b(m) \right\} e^{in\theta}$$

we set $k = n - m$ to obtain

$$\sum_k \left\{ \sum_m a(k)b(m) \right\} e^{i(k+m)\theta} = \sum_k e^{ik\theta} \left\{ \sum_m a(k)b(m)e^{im\theta} \right\}$$

$$= \left\{ \sum_k a(k)e^{ik\theta} \right\} \left\{ \sum_m b(m)e^{im\theta} \right\} = \widehat{a}(\theta) \cdot \widehat{b}(\theta)$$

To see how all of this might be useful in the study of Fourier series, consider the following result of N. Wiener: Suppose that $f \in C(T)$ is never zero, and the Fourier series of f is absolutely convergent. Then the Fourier series of $\frac{1}{f}$ is absolutely convergent.

As one might expect, a proof of this result using classical methods is quite difficult. The reader may want to try to find one. Our discussion of the Fourier transform, however, enables us to reformulate the theorem as follows:

It is clear that a function in $C(T)$ is in the range of $\Im$ if, and only if, its Fourier series is absolutely convergent. Thus the hypothesis of Wiener's theorem tells us that there is an $a \in \ell^1(Z)$ such that $\widehat{a} = f$. Suppose now that we can find $b \in \ell^1(Z)$ for which $a * b = e$. Then $\Im(a*b) = \widehat{a} \cdot \widehat{b} = \Im(e) = 1$; i.e., $\widehat{a}(\theta) \cdot \widehat{b}(\theta) = f(\theta) \cdot \widehat{b}(\theta) = 1$ for all θ. Because $f(\theta)$ is never zero, this says that $\widehat{b} = \frac{1}{f}$. Thus $\frac{1}{f}$ is in the range of $\Im$, it is equal to $\widehat{b}$, and hence its Fourier series is absolutely convergent.

In other words Wiener's theorem, a theorem about Fourier series, is equivalent to the following: If $a \in \ell^1(Z)$ and $\widehat{a}$ is never zero, then there is a $b \in \ell^1(Z)$ such that $a * b = e$; i.e., if $a \in \ell^1(Z)$ and $\widehat{a}$ never vanishes, then a has a "multiplicative inverse" in ℓ^1. We shall investigate this result more fully in the next section. To begin, let us define our terms.

Definition 2. An element a of $\ell^1(Z)$ is said to be invertible if there is an element b in $\ell^1(Z)$ such that $a * b = b * a = e$. We call b the (see Exercises 1, problem 2(e)) inverse of a and denote it by a^{-1}.

It is clear that if $a \in \ell^1$ has an inverse, then $\widehat{a}$ is never zero, because $\widehat{a * a^{-1}} = \widehat{a}(\widehat{a^{-1}}) = \widehat{e} = 1$. What is difficult is showing that when $\widehat{a}$ is never zero, a has an inverse.

We end this section with a proof that $(\ell^1(Z), \|\cdot\|_1)$ is a Banach space; i.e., every Cauchy sequence of points in this space converges to a point of this space (Chapter Three, Section 2, Definition 3).

Lemma 5. The space $(\ell^1(Z), \|\cdot\|_1)$ is a complete normed space.

Proof. Let $\{a^n\}_{n=1}^\infty$ be a Cauchy sequence in $\ell^1(Z)$ and, given $\varepsilon > 0$, choose N so that
$$\|a^n - a^m\|_1 < \varepsilon$$
whenever both m and n exceed N. Then for any fixed j_0 and any fixed k with $j_0 < k$, we have
$$|a^n(j_0) - a^m(j_0)| \leq \sum_{j=-k}^{k} |a^n(j) - a^m(j)| \leq \sum_{-\infty}^{\infty} |a^n(j) - a^m(j)| < \varepsilon$$
whenever both m and n exceed N. It follows that $\{a^n(j)\}_{n=1}^\infty$ is a Cauchy sequence in $\mathbb{K}$ for each fixed j. Because $\mathbb{K}$ is complete, this sequence converges.

Let $b(j) = \lim_{n \to \infty} a^n(j)$ for each $j \in Z$. Now things get a little bit messy. We let

$$b_k = \{\cdots, 0, b(-k), b(-k+1), \cdots, b(0), b(1), \cdots, b(k), 0, \cdots\}$$

and we let, for each n,

$$b_k^n = \{\cdots, 0, a^n(-k), \cdots, a^n(0), \cdots, a^n(k), 0, \cdots\}$$

Then b_k and b_k^n are in $\ell^1(Z)$ and we have

$$\|b_k\|_1 \leq \|b_k - b_k^n\|_1 + \|b_k^n\|_1 \leq \varepsilon + \sum_{j=-k}^{k} |a^n(j)|$$

whenever $n \geq N$. If we fix $n \geq N$ and let k tend to infinity we find that

$$\sum_{-\infty}^{\infty} |b(j)| \leq \varepsilon + \|a^n\|_1$$

showing that $b = \{b(j)\}_{-\infty}^{\infty}$ is in $\ell^1(Z)$.

Finally, we shall show that $\lim_{n \to \infty} \|a^n - b\|_1 = 0$. As we saw above, for any fixed k,

$$\sum_{j=-k}^{k} |a^n(j) - a^m(j)| < \varepsilon$$

whenever both m and n exceed N. Letting m go to infinity we get

$$\sum_{j=-k}^{k} |a^n(j) - b(j)| \leq \varepsilon$$

whenever $n \geq N$. Now choose, and fix, $n \geq N$ and let k go to infinity. We find

$$\|a^n - b\|_1 \leq \varepsilon$$

whenever $n \geq N$.

Exercises 1

1. Let $a \in \ell^1(Z)$ and, for a fixed $m \in Z$, let $a_m(n) = a(n - m)$ for all n. Show that $\widehat{a}_m(\theta) = e^{+im\theta}\widehat{a}(\theta)$.

*2. For any a, b, c in $\ell^1(\mathbb{Z})$ and any $\lambda \in \mathbb{C}$ show that:

   (a) $a * b = b * a$;

   (b) $\lambda(a * b) = (\lambda a) * b = a * (\lambda b)$;

   (c) $a * (b * c) = (a * b) * c$;

   (d) $a * (b + c) = a * b + a * c$.

   (e) Show that the inverse of a, when it exists, is unique.

*3. Let $\{b_n\} \subseteq \ell^1$ and suppose that $\lim b_n = b_0$ for the ℓ^1-norm. Show that $\lim a * b_n = a * b_0$ for any $a \in \ell^1$. If $a \in \ell^1$ is fixed, show that the map $T_a : \ell^1 \to \ell^1$ defined by $T_a(b) = a * b$ is linear and continuous.

4. Let $c \in \ell^1$ be defined as follows: $c(n) = 1$ if $n = 0$ or $n = 1$, and $c(n) = 0$ for all other n.

   (a) Show that $(c * a)(n) = a(n) + a(n-1)$, here $a \in \ell^1$ is arbitrary.

   (b) Let $c^1 = c$, $c^2 = c * c$, ... Show that for any fixed $m > 0$
   $$c^m(n) = \binom{m}{n} = \frac{m!}{n!(m-n)!}.$$

5. Define $g \in \ell^1(\mathbb{Z})$ by letting $g(1) = 1$, $g(n) = 0$ for $n \neq 1$.

   *(a) Let $g^1 = g$, $g^2 = g * g$, ... and show that for k fixed, $g^k(n) = 1$ when $n = k$ and $g^k(n) = 0$ when $n \neq k$.

   *(b) Let $g^{-1}(n) = 1$ when $n = -1$, $g^{-1}(n) = 0$ for $n \neq 1$. Show that g^{-1} is the inverse of g.

   *(c) Compute $(g^{-1})^k$ for each positive integer k.

   *(d) For any fixed $a \in \ell^1$ show that $a = \sum_{-\infty}^{\infty} a(n) g^n$ where convergence is for the ℓ^1-norm; here $g^0 = e$.

   (e) Given a, b in ℓ^1 use the series representation of these elements given in (d) to find a series representation for $a * b$.

   (f) Use the representation given in (d) to show that to every continuous linear map $\varphi : \ell^1(\mathbb{Z}) \to \mathbb{C}$ there corresponds a unique sequence $\{c_n\}$ such that

   i. $\varphi(a) = \sum a_n c_n$ for all $a \in \ell^1$;

   ii. $\sup\{|c_n| \mid n \in \mathbb{Z}\}$ is bounded.

6. If $f \in C(T)$ is never zero, show that $\frac{1}{f}$ is in $C(T)$.

2. Invertible Elements in $\ell^1(Z)$

We recall that $e(n) = 0$ when $n \neq 0$, $e(0) = 1$, and an element a of ℓ^1 is invertible if there is an element $a^{-1} \in \ell^1$ such that $a * a^{-1} = a^{-1} * a = e$.

For $a \in \ell^1$, $a \neq 0$, we set $a^0 = e$, $a^1 = a$ and, for any integer $k > 1$, we set $a^k = a * a^{k-1}$.

Theorem 1. Let $a \in \ell^1(Z)$ and suppose that $\|a\|_1 < 1$. Then the element $(e - a)$ is invertible. Furthermore, if $a \neq 0$,

$$(e-a)^{-1} = \sum_{j=0}^{\infty} a^j$$

where convergence is for the ℓ^1-norm.

Proof. For each $n \geq 1$ let us set $b_n = \sum_{j=0}^{n} a^j$ and consider the sequence $\{b_n\} \subseteq \ell^1$. When $m > n$ we have

$$\|b_m - b_n\|_1 = \|\sum_{j=n+1}^{m} a^j\|_1 \leq \sum_{j=n+1}^{m} \|a^j\|_1$$

$$\leq \sum_{n+1}^{m} \|a\|_1^j \quad \text{(Lemma 3 of Section 1 of this chapter)}$$

$$\leq \sum_{n+1}^{\infty} \|a\|_1^j \leq \frac{\|a\|_1^{n+1}}{1 - \|a\|_1}$$

because the series of real numbers $\sum_0^{\infty} \|a\|_1^j$ is geometric, and because $\|a\|_1 < 1$, it converges to $\frac{1}{1-\|a\|_1}$. But, again because $\|a\|_1 < 1$, $\lim_{n \to \infty} \|a\|_1^{n+1} = 0$ showing that $\{b_n\}$ is a Cauchy sequence in $(\ell^1, \|\cdot\|_1)$. By Lemma 5 of Section 1, this sequence converges to an element of ℓ^1 that we shall denote by

$$\sum_{j=0}^{\infty} a^j$$

We shall now show that this sum is the inverse of $(e - a)$. We have

$$(e-a) * \left(\sum_{j=0}^{\infty} a^j\right) = (e-a) * \left(\lim_{n \to \infty} \sum_{0}^{n} a^j\right)$$

$$= \lim_{n \to \infty} \left[(e-a) * \sum_{0}^{n} a^j\right] \quad \text{(Exercises 1, problem 3)}$$

$$= \lim_{n \to \infty} \left[\sum_{j=0}^{n} a^j - \sum_{j=0}^{n} a^{j+1}\right] \lim_{n \to \infty} (e - a^{n+1}) = e$$

By Exercises 1, problem 2(a), reversing the factors give the same result.

Corollary 1. Let $a \in \ell^1(Z)$ be invertible and suppose that $b \in \ell^1$ satisfies $\|a - b\|_1 < \frac{1}{\|a^{-1}\|_1}$. Then b is invertible.

Proof. We have $\|e - a^{-1} * b\|_1 = \|a^{-1} * (a-b)\|_1 \leq \|a^{-1}\|_1 \|a-b\|_1 < 1$ by Lemma 3 of Section 1 and our hypothesis. Thus, by Theorem 1, $[e - (e - a^{-1} * b)] = a^{-1} * b$ is invertible. Call this c. So $c = a^{-1} * b$ and c is invertible. But then $b = a * c$, and this last convolution is invertible (its inverse is $c^{-1} * a^{-1}$).

It follows from Corollary 1 that the set of all invertible elements of ℓ^1 is open and hence that its complement, what is sometimes called the set of singular elements of ℓ^1, is a closed set.

Definition 1. Recall that a linear map from $\ell^1(Z)$ into $\mathbb{C}$ is called a linear functional on $\ell^1(Z)$. A linear functional φ, that is not identically zero, is called a multiplicative functional if $\varphi(a * b) = \varphi(a) \cdot \varphi(b)$ for all a, b in $\ell^1(Z)$.

If φ is a multiplicative functional, then $\varphi(a) \neq 0$ for some a. Consequently, $\varphi(a) = \varphi(a * e) = \varphi(a) \cdot \varphi(e)$, and we see that $\varphi(e)$ must be one. Furthermore, if b is invertible, then $\varphi(b) \neq 0$ and, in fact, $\varphi(b^{-1}) = \frac{1}{\varphi(b)}$. Theorem 1 tells us more about these functionals.

Corollary 2. If φ is a multiplicative functional on $\ell^1(Z)$, then for every a in this space we have $|\varphi(a)| \leq \|a\|_1$. In particular, every multiplicative functional on $\ell^1(Z)$ is continuous on this space (Chapter Three, Section 2, Theorem 1).

Proof. We argue by contradiction. Suppose that for some a we have $|\varphi(a)| > \|a\|_1$. Set $\lambda = \varphi(a)$ and note that because $\left\|\frac{a}{\lambda}\right\|_1 < 1$, Theorem 1 tells us that $\left(e - \frac{a}{\lambda}\right) \neq 0$ is invertible. But then $\varphi\left(e - \frac{a}{\lambda}\right) \neq 0$ showing that because φ is a linear map, $\varphi\left(\frac{a}{\lambda}\right) \neq 1$. This contradicts the definition of λ.

Let us now choose, and fix, a multiplicative functional φ on ℓ^1 and consider $M(\varphi) = \{a \in \ell^1 \mid \varphi(a) = 0\}$. The properties of $M(\varphi)$ are easily proved but worth listing.

(a) $M(\varphi)$ is closed.

   Suppose $\{a_n\} \subseteq M(\varphi)$ and $\lim a_n = a_0$. Then $\lim \varphi(a_n) = \varphi(a_0)$, and clearly this gives us $\varphi(a_0) = 0$; i.e., $a_0 \in M(\varphi)$.

(b) $M(\varphi)$ is a linear subspace of ℓ^1, $M(\varphi) \neq \ell^1$, and if N is a linear subspace for which $M(\varphi) \subseteq N$, then either $M(\varphi) = N$ or $N = \ell^1$.

Because φ is linear, it is clear that $M(\varphi)$ is a linear subspace of ℓ^1, and because φ is not identically zero, $M(\varphi) \neq \ell^1$; in fact, we may choose $a_0 \in \ell^1$ such that $\varphi(a_0) \neq 0$. Now given $b \in \ell^1$ set $\alpha = \frac{\varphi(b)}{\varphi(a_0)}$. Clearly $\varphi(b - \alpha a_0) = 0$ so $b - \alpha a_0 = c \in M(\varphi)$, and we see that $b = c + \alpha a_0$. Now suppose $M(\varphi) \subseteq N$. If $M(\varphi) = N$, we are done. If there is a $b \in N \setminus M(\varphi)$, then $b = c + \alpha a_0$ for some scalar α and some $c \in M$. Because $\alpha \neq 0$ and N is a linear subspace, we see that $a_0 \in N$. But then $N = \ell^1$ as claimed.

(c) If $a \in M(\varphi)$ and $b \in \ell^1$, then $b * a \in M(\varphi)$.

Because $\varphi(b * a) = \varphi(b) \cdot \varphi(a)$ this is obvious.

Definition 2. A nonempty subset I of $\ell^1(Z)$ is said to be an ideal in $\ell^1(Z)$ if (a) I is a linear subspace; (b) whenever $a \in I$ and $b \in \ell^1(Z)$, $b * a$ is in I. An ideal I is called a proper ideal if $I \neq \ell^1(Z)$. Finally, a proper ideal is called a maximal ideal if the only ideals that contain it are itself and $\ell^1(Z)$.

We observe that a proper ideal cannot contain e nor can it contain any invertible element. Thus if I is a proper ideal, then I is contained in the set of singular elements of ℓ^1 (defined just before Definition 1). Because this last set is closed, we see that the closure (Chapter 3, Exercises 2, problem 2) of I is contained in it. We leave it to the reader to show that the closure of an ideal is an ideal and we have shown:

Lemma 1. The closure of a proper ideal in ℓ^1 is a proper ideal in ℓ^1. In particular, every maximal ideal in ℓ^1 is closed.

Let $\mathcal{A}$ be a nonempty family of nonempty sets. A subfamily $\mathcal{N}$ of $\mathcal{A}$ is said to be a nest if for any sets A, B in $\mathcal{N}$ either $A \subseteq B$ or $B \subseteq A$. A nest $\mathcal{M}$ is called a maximal nest if the conditions (i) $\mathcal{N}$ is a nest and (ii) $\mathcal{M} \subseteq \mathcal{N}$ together imply that $\mathcal{M} = \mathcal{N}$. So a maximal nest $\mathcal{M}$ is as large as possible; more precisely, if $A \notin \mathcal{M}$, then there must be a set $B \in \mathcal{M}$ such that neither $A \subseteq B$ nor $B \subseteq A$ is true.

There is a useful equivalent form of the axiom of choice (Chapter Zero, Section 2) that was first proved by Hausdorff. Maximum Principle: Given any nest $\mathcal{N}$ in $\mathcal{A}$ there is a maximal nest $\mathcal{M}$ in $\mathcal{A}$ such that $\mathcal{N} \subseteq \mathcal{M}$.

Lemma 2. Every proper ideal in $\ell^1(Z)$ is contained in a maximal ideal.

Proof. Let I be a proper ideal in $\ell^1(Z)$ and let $\mathcal{I}$ be the family of all proper ideals in this set. Then $\{I\}$ is a nest in $\mathcal{I}$. By the maximum principle there is a maximal nest $\mathcal{M}$ in $\mathcal{I}$ that contains $\{I\}$. It is convenient to index the ideals in $\mathcal{M}$ (Chapter Zero, Section 2, just after Definition 3). So we may write $\mathcal{M} = \{I_\alpha \mid \alpha \in A\}$, and we know that because $\{I\} \subseteq \mathcal{M}$, $I = I_\alpha$ for some $\alpha \in A$. Now let $J = \bigcup_{\alpha \in A} I_\alpha$ (Chapter Zero, Section 2, just after Definition 3). It is clear that $I \subseteq J$. We shall show that J is a maximal ideal.

Our first task is to show that J is a linear subspace of $\ell^1(Z)$. So given a, b in J we must show that any linear combination of these two vectors is in J. But we have $a \in I_\alpha$ and $b \in I_\beta$ for α, β in A. Because $\mathcal{M}$ is a nest, either $I_\alpha \subseteq I_\beta$ or $I_\beta \subseteq I_\alpha$. Let us suppose the former. Then a and b are in I_β, and because I_β is a linear subspace of $\ell^1(Z)$, any linear combination of a and b are in I_β. But $I_\beta \subseteq J$ so we see that J is a linear subspace of $\ell^1(Z)$.

Next we must show that for any $a \in J$ and any $c \in \ell^1(Z)$, $a * c$ is in J. Now $a \in I_\alpha$ for some α and because I_α is an ideal $a * c$ is in this set. Thus $a * c \in I_\alpha \subseteq J$; and we have shown that J is an ideal in $\ell^1(Z)$ that contains I.

Now $J = \bigcup_{\alpha \in A} I_\alpha$ and each I_α is a proper ideal. Thus $e \notin I_\alpha$ all $\alpha \in A$, and so $e \notin J$ and we see that J is a proper ideal.

Finally, we must show that J is a maximal ideal. Suppose J' is a proper ideal in $\ell^1(Z)$ and suppose that $J \subsetneq J'$. Then $\mathcal{M} \cup \{J'\}$ is a nest in $\mathcal{I}$ and $\mathcal{M} \subseteq \mathcal{M} \cup \{J'\}$. Because $\mathcal{M}$ is a maximal nest, we must have $\mathcal{M} = \mathcal{M} \cup \{J'\}$; thus $J' \in \mathcal{M}$ and so $J' = I_\gamma$ for some $\gamma \in A$. Clearly, we have reached a contradiction because $I_\gamma \subseteq J$.

Our discussion above, see (b) and (c), shows that the null space of any multiplicative functional is a maximal ideal in $\ell^1(Z)$. Using some abstract algebra and some complex function theory (Liouville's theorem), one can show that the converse is true. Because we are not supposing that our readers are familiar with these topics we shall only state:

(*) Any proper closed ideal of $\ell^1(Z)$ is contained in the null space of a multiplicative functional on $\ell^1(Z)$.

Theorem 2. Given a maximal ideal M in $\ell^1(Z)$, there is a $\theta \in T$ such that $M = \{a \in \ell^1(Z) \mid \widehat{a}(\theta) = 0\}$.

Proof. Let φ be a multiplicative functional on ℓ^1 such that $M = \{a \in \ell^1 \mid \varphi(a) = 0\}$. Such a φ exists by (*). Now recall the element g defined in Exercises 1, problem 5: $g(1) = 1$, $g(n) = 0$ for $n \neq 1$. Because $|\varphi(a)| \leq \|a\|_1$ for every $a \in \ell^1$ and because $\|g\|_1 = 1$, we see that $\lambda = \varphi(g)$ must satisfy $|\lambda| \leq 1$. But $\|g^{-1}\|_1 = 1$ so $|\frac{1}{\lambda}| \leq 1$ also and we conclude that $\lambda = \varphi(g) = e^{+i\theta}$ for some fixed $\theta \in T$.

Now for any $a \in \ell^1$ we have

$$\varphi(a) = \varphi\left(\lim_{k \to \infty}\left[\sum_{-k}^{k} a(n)g^n\right]\right) = \lim_{k \to \infty} \varphi\left(\sum_{-k}^{k} a(n)g^n\right)$$

$$= \lim_{k \to \infty} \sum_{-k}^{k} a(n)\varphi(g)^n = \sum_{-\infty}^{\infty} a(n)e^{in\theta}$$

Thus $\varphi(a)$ is just $\widehat{a}$ evaluated at θ, and so $M = \{a \mid \varphi(a) = 0\} = \{a \in \ell^1 \mid \widehat{a}(\theta) = 0\}$ as claimed.

Theorem (N. Wiener). If $f \in C(T)$ is never zero and if the Fourier series of f converges absolutely, then the Fourier series of $\frac{1}{f}$ converges absolutely.

Proof. We cannot claim to have here a complete proof because we simply stated (*) and it is crucial. We are only giving a sketch. As we have seen we can choose $b \in \ell^1$ such that $\hat{b} = f$ and our hypothesis tells us that $\hat{b}$ is never zero on T. Suppose that b does *not* have an inverse in ℓ^1, and let $I = \{a * b \mid a \in \ell^1\}$. Clearly, I is an ideal and because $e \notin I$, it is a proper ideal in ℓ^1. Thus there is a multiplicative functional φ such that $I \subseteq M = \{a \in \ell^1 \mid \varphi(a) = 0\}$. Then for some $\theta \in T$ we must have $M = \{a \in \ell^1 \mid \hat{a}(\theta) = 0\}$. But this says $\hat{b}(\theta) = 0$, which is a contradiction. It follows that b is invertible in ℓ^1; i.e., there is a $b^{-1} \in \ell^1$ such that $b^{-1} * b = e$. Clearly, $\widehat{(b^{-1} * b)} = \hat{e}$ or $\widehat{(b^{-1})} \cdot f = 1$.

Exercises 2

* 1. For any fixed $a \in \ell^1(Z)$, let $I(a) = \{a * b \mid b \in \ell^1(Z)\}$.

    (a) Show that $I(a)$ is an ideal in $\ell^1(Z)$ that contains a.

    (b) Show that $I(a)$ is a proper ideal if, and only if, a is *not* invertible.

    (c) Show that the closure of a proper ideal is a proper ideal.

2. For any $a \in \ell^1(Z)$, let $\sigma(a) = \{\lambda \in \mathbb{C} \mid (a - \lambda e) \text{ is not invertible}\}$. We call $\sigma(a)$ the spectrum of a, and we call $\mathbb{C} \setminus \sigma(a) \equiv \rho(a)$ the resolvent set of a.

    (a) Show that $\sigma(a)$ is a closed set by showing that $\rho(a)$ is an open set.

    (b) Show that any $\lambda \in \mathbb{C}$ such that $|\lambda| > \|a\|_1$ is in $\rho(a)$ and hence $\sigma(a) \subseteq \{\lambda \in \mathbb{C} \mid |\lambda| \leq \|a\|_1\}$.

3. The Fourier Transform on $\mathbb{R}$

The theory here is very much like that for the case of Z except that the definitions, and the proofs of the theorems, involve subtleties not found in the discrete case. Let us begin informally, pursuing, as much as possible, the analogy with $\ell^1(Z)$. Because this space consists of all $a : Z \to \mathbb{C}$ such that $\sum_{n=-\infty}^{\infty} |a(n)| < \infty$; we "should" define $L^1(\mathbb{R})$ to be

$$\{f : \mathbb{R} \to \mathbb{C} \mid \int_{-\infty}^{\infty} |f(t)| dt < \infty\}$$

Assuming that this is a vector space we set, for all $f \in L^1(\mathbb{R})$,

$$\|f\|_1 = \int_{-\infty}^{\infty} |f(t)| dt$$

With a "proper" interpretation of the integral, we can show that $(L^1(\mathbb{R}), \|\cdot\|_1)$ is a Banach space. (See Chapter Three, Exercises 5, problem 3(c) for a construction of this space.) The integral needed is that of Lebesgue, an extension of the classical integral of Riemann. We shall not assume a knowledge of this integral and so our proofs of the theorems stated below are only "sketches"; i.e., they are incomplete.

For each $f \in L^1(\mathbb{R})$ we define

$$\widehat{f}(x) = \int_{-\infty}^{\infty} f(t)e^{ixt}dt$$

Observe that $\widehat{f}$ is also defined for all real numbers. This is a problem, and it is wise to distinguish the domain of f, which is $\mathbb{R}$, from that of $\widehat{f}$. We shall do so by agreeing that $\widehat{f}$ is defined on a "second" real line that we shall denote by $\widehat{\mathbb{R}}$.

For each $a \in \ell^1(\mathbb{Z})$, $\widehat{a}$ is a continuous function on the unit circle; i.e., $\widehat{a}$ is an element of $C(T)$. Let $C(\widehat{\mathbb{R}})$ be the vector space of all continuous functions on $\widehat{\mathbb{R}}$. Then we have:

Lemma 1. For each $f \in L^1(\mathbb{R})$, $\widehat{f} \in C(\widehat{\mathbb{R}})$. Moreover, the map $\Im$ that takes each f to $\widehat{f}$ is linear and one to one.

Proof. Let $f \in L^1(\mathbb{R})$ be given. Then

$$\widehat{f}(x+h) = \int_{-\infty}^{\infty} f(t)e^{i(x+h)t}dt = \int_{-\infty}^{\infty} e^{iht}f(t)e^{ixt}dt$$

Hence

$$\widehat{f}(x+h) - \widehat{f}(x) = \int_{-\infty}^{\infty} (e^{iht} - 1)e^{ixt}f(t)dt$$

and

$$|\widehat{f}(x+h) - \widehat{f}(x)| \leq \int_{-\infty}^{\infty} |f(t)||e^{iht} - 1|dt$$

As $h \to 0$ so does $|e^{iht} - 1|$ and, one can show, so does the integral. Thus $\widehat{f} \in C(\mathbb{R})$.

It is easy to see that $\Im : L^1(\mathbb{R}) \to C(\mathbb{R})$ is a linear map. To show that $\Im$ is one-to-one, we need only show that one can recapture $f \in L^1(\mathbb{R})$ from a knowledge of its Fourier transform $\widehat{f}$. To do this one uses a kind of Cesáro convergence for integrals. We have

$$f(t) = \lim_{R \to \infty} \frac{1}{2\pi} \int_{-R}^{R} \left(1 - \frac{|x|}{R}\right) \widehat{f}(x) e^{-ixt} dx$$

where convergence is for the L^1-norm (compare this to Chapter One, Exercises 6, problem 3).

Note. When $\widehat{f}$ happens to be integrable this formula gives

$$f(t) = \frac{1}{2\pi} \int_{-\infty}^{\infty} \widehat{f}(x) e^{-ixt} dx$$

which is valid in most practical applications.

Lemma 2. For any $f \in L^1(\mathbb{R})$, $\lim_{|x|\to\infty} \widehat{f}(x) = 0$. Furthermore,

$$\max_{x\in\mathbb{R}} |\Im(f)(x)| = \max_{x\in\mathbb{R}} |\widehat{f}(x)| \leq \|f\|_1$$

Proof. The first statement is called the Riemann-Lebesgue theorem. It generalizes the theorem of Riemann given in Section 5 of Chapter 1 (Theorem 1). We have

$$\widehat{f}(x) = \int f(t) e^{ixt} dt \quad \text{and} \quad -\widehat{f}(x) = \int_{-\infty}^{\infty} e^{ix(t+\frac{\pi}{x})} f(t) dt = \int_{-\infty}^{\infty} e^{ixt} f(t - \frac{\pi}{x}) dt$$

Hence

$$2|\widehat{f}(x)| \leq \int_{-\infty}^{\infty} |f(t) - f(t - \frac{\pi}{x})| dt$$

For any $f \in L^1(\mathbb{R})$ this last integral tends to zero a $|x| \to \infty$.
We leave the second statement to the reader.

Let us illustrate our results by calculating the Fourier transform of some specific functions. In these examples a is a fixed, positive, real number.

(a) Let $U(t) = 1$ when $t \geq 0$, $U(t) = 0$ when $t < 0$. Set $f(t) = e^{-at} U(t)$.

$$\int_{-\infty}^{\infty} |f(t)| dt = \int_0^{\infty} e^{-at} dt = \frac{1}{a}$$

so f is in $L^1(\mathbb{R})$. Now

$$\widehat{f}(x) = \int_{-\infty}^{\infty} e^{-at} U(t) e^{ixt} dt = \int_0^{\infty} e^{(-a+ix)t} dt$$

$$= \lim_{M\to\infty} \left[\frac{e^{(-a+ix)M}}{-a+ix} - \frac{1}{-a+ix} \right] = \frac{1}{a-ix}$$

Note that $\widehat{f}$ is continuous and that $\lim_{|x|\to\infty} |\widehat{f}(x)| = \lim \frac{1}{\sqrt{a^2+x^2}} = 0$.

(b) Let $1_a(t) = 1$ for $|t| \leq a$, $1_a(t) = 0$ for $|t| > a$.

$$\widehat{1_a}(x) = \int_{-a}^{a} e^{ixt} dt = \frac{e^{iax} - e^{-iax}}{ix} = \frac{2\sin ax}{x}$$

Again we note that the Fourier transform is continuous and tends to zero as $|x| \to \infty$.

(c) Referring to example (b), let $k(t) = (1 - |t|)1_1(t)$. Because $k(-t) = k(t)$ we see that

$$\widehat{k}(x) = \int_{-1}^{1} k(t) \cos xt \, dt = 2 \int_{0}^{1} (1-t) \cos xt \, dt$$

$$= 2 \left[\frac{1 - \cos x}{x^2} \right] = \frac{4}{x^2} \sin^2\left(\frac{x}{2}\right) = \left[\frac{\sin(\frac{x}{2})}{x/2} \right]^2$$

(d) Consider the Abel kernel $f(t) = e^{-a|t|}$.

$$\widehat{f}(x) = \int_{-\infty}^{\infty} e^{-a|t|} e^{ixt} dt = \int_{-\infty}^{0} e^{(a+ix)t} dt + \int_{0}^{\infty} e^{(-a+ix)t} dt$$

$$= \lim_{M \to -\infty} \left[\frac{1}{a + ix} - \frac{e^{(a+ix)M}}{a+ix} \right] + \lim_{N \to \infty} \left[\frac{e^{(-a+ix)N}}{-a+ix} - \left(\frac{1}{-a+ix} \right) \right]$$

$$= \frac{2a}{a^2 + x^2}$$

(e) Here we simply note that when $\widehat{f}$ is differentiable we have

$$\frac{d}{dx} \widehat{f}(x) = \frac{d}{dx} \int_{-\infty}^{\infty} f(t) e^{ixt} dt = \int_{-\infty}^{\infty} [it \, f(t)] e^{ixt} dt$$

which is the Fourier transform of $it\, f(t)$. More generally, $\frac{d^{(n)}}{dx^n} \widehat{f}(x)$ is the Fourier transform of $(it)^n f(t)$.

In the space $\ell^1(Z)$ we found a kind of multiplication. There is an analogous operation here.

Definition 1. For any $f, g \in L^1(\mathbb{R})$ we define

$$(f * g)(x) = \int_{-\infty}^{\infty} f(x - y) g(y) dy$$

Lemma 3. For any f, g in $L^1(\mathbb{R})$ the function $f * g$ is in $L^1(\mathbb{R})$ and $\|f * g\|_1 \leq \|f\|_1 \|g\|_1$.

Proof.

$$\|f * g\|_1 = \int |(f * g)(x)| dx = \int |\int f(x-y)g(y)dy| dx$$
$$\leq \int dx \int |f(x-y)||g(y)| dy = \int |g(y)| dy \int |f(x-y)| dx$$
$$= \|g\|_1 \int_{-\infty}^{\infty} |f(x-y)| dx = \|g\|_1 \|f\|_1$$

assuming the interchange in the order of integration is valid.

Lemma 4. For any f, g in $L^1(\mathbb{R})$ we have $\Im(f * g) = \Im(f)\Im(g)$ or, in another notation, $\widehat{(f * g)} = \hat{f} \cdot \hat{g}$.

Proof. $\widehat{(f * g)}(x) = \int (f * g)(t) e^{ixt} dt = \int e^{ixt} \left(\int f(t-u)g(u) du \right) dt$ setting $w = t - u$ this becomes

$$\int e^{ix(w+u)} du \int f(w)g(u) dw = \int e^{ixu} g(u) du \int e^{ixw} f(w) dw$$
$$= \hat{g}(x) \cdot \hat{f}(x)$$

A typical application of the Fourier transform is to solve differential equations. As an illustration of this consider

$$-\frac{d^2 u}{dt^2} + a^2 u(t) = f(t)$$

where $f(t)$ is assumed known. Applying $\Im$ we get

$$\Im\left[-\frac{d^2 u}{dt^2} + a^2 u \right] = -(ix)^2 \hat{u} + a^2 \hat{u} = (x^2 + a^2)\hat{u} = \hat{f}$$

Thus

$$\hat{u}(x) = \left(\frac{1}{x^2 + a^2} \right) \hat{f}(x) = \hat{g}(x) \hat{f}(x)$$

where, from (d),

$$g(t) = \frac{1}{2a} e^{-a|t|}$$

It follows, then, from Lemma 4 that

$$u(t) = \frac{e^{-a|t|}}{2a} * f(t) = \frac{1}{2a} \int_{-\infty}^{\infty} e^{-a|t-s|} f(s) ds$$

Exercises 3

1. Let $f(t) = e^{2it} 1_1(t)$. Compute $\widehat{f}(x)$.

2. Let $f \in L^1(\mathbb{R})$ and let a, h be real numbers. Show that

   (a) For $a \neq 0$, $\Im[f(at)] = \frac{1}{|a|}\widehat{f}(\frac{x}{a})$;

   (b) $\Im[f(t-h)] = e^{+ixh}\widehat{f}(x)$;

   (c) $\Im[e^{iat}f(t)] = \widehat{f}(x+a)$;

   (d) $\Im[(\cos at)f(t)] = \frac{1}{2}\widehat{f}(x-a) + \frac{1}{2}\widehat{f}(x+a)$.

3. For any f, g, h in $L^1(\mathbb{R})$ and any $\lambda \in \mathbb{C}$ show that

   (a) $f * g = g * f$;

   (b) $\lambda(f * g) = (\lambda f) * g = f * (\lambda g)$;

   (c) $f * (g * h) = (f * g) * h$;

   (d) $f * (g + h) = f * g + f * h$.

*4. Show that there is no function $e(t) \in L^1(\mathbb{R})$ such that $e * f = f * e = f$ for all $f \in L^1(\mathbb{R})$. Hint: If e exists, then $e * e = e$. Now apply the Fourier transform.

4. Naive Group Theory

We can get a deeper understanding of the Fourier transform by using some ideas from modern algebra. These are usually discussed abstractly, but, we believe, it may be better to begin with some concrete examples.

Definition 1. A nonempty subset A of $\mathbb{C}$ is said to be an additive group if $(a - b) \epsilon A$ whenever both a and b are in A.

It is clear that $\mathbb{C}, \mathbb{R}, Q, Z$ and even $\{0\}$ are additive groups. The last of these is called the trivial group. We might also recall that, for any function $f : \mathbb{R} \to \mathbb{K}$, the set $P(f)$ is an additive group (Chapter One, Section 2, Lemma 1). It is a nontrivial group when, and only when, the function f is periodic. Many of the properties of $P(f)$ are shared by any additive group.

Lemma 1. Let A be an additive group. Then (i) $0 \epsilon A$; (ii) for any $a \epsilon A$ the number $(-a)$ is in A; (iii) when a and b are in A so is $(a + b)$; (iv) for any $a \epsilon A$ the set $\{na \mid n \epsilon Z\}$ is contained in A.

Proof. We may assume that A is nontrivial. Then for $a \epsilon A$ we must have $(a - a)$ in A, which proves (i). Now that we know $0 \epsilon A$, $(0 - a)$ is in A for every $a \epsilon A$, proving (ii). To prove (iii) we note that if a, b are in A, then $a, -b$ are in A, and so $a - (-b)$ is in A.

Finally, let $a \epsilon A$ and suppose that na is *not* in A for some positive integer n. Let n_0 be the smallest positive integer such that $n_0 a \notin A$. Then by (iii) we see that $a + a = 2a \epsilon A$; hence $n_0 > 2$. It follows that $(n_0 - 1) a \epsilon A$ because of the way n_0 was defined. But, then, again using (iii), $(n_0 - 1) a + a = n_0 a$ is in A, which is a contradiction. We have shown that $na \epsilon A$ for every positive integer n. By (i), however, $0 \cdot a$ is in A, and by (ii) $(-na) = (-n) a$ is in A for every n. Thus $\{na \mid n \epsilon Z\} \subseteq A$ as claimed.

When A, B are additive groups and $B \subseteq A$ we shall say that B is a subgroup of A. In particular, when $a \epsilon A$ the set $\{na \mid n \epsilon Z\}$, which is clearly an additive group, is a subgroup of A.

Clearly, every additive group is a subgroup of $\mathbb{C}$, but it is often useful to specify that we are working with a subgroup of $\mathbb{R}$ or a subgroup of Z.

Corollary 1. Let A be a subgroup of the additive group $\mathbb{R}$. If A contains a smallest positive member, say a_0, then $A = \{na_0 \mid n \epsilon Z\}$.

Proof. We have seen that $\{na_0 \mid n \epsilon Z\} \subseteq A$. Given $b \epsilon A$ suppose, first, that b is positive. Then $a_0 \leq b$, and we may write $b = ka_0 + r$ where k is a positive integer and $0 \leq r < a_0$. Because both b and ka_0 are in A, it follows that $r \epsilon A$ and hence that $r = 0$. Thus $b \epsilon \{na_0 \mid n \epsilon Z\}$. When b is negative, $(-b)$ is a positive member of A and so $(-b) = ka_0$ for some integer k. But then $b = (-k) a_0$ and we are done.

Corollary 2. If A is a subgroup of the additive group Z, then either $A = \{0\}$ or $A = \{n\ell \mid n \epsilon Z\}$ where ℓ is the smallest positive member of A.

Definition 2. A nonempty subset M of $\mathbb{C}$ is said to be a multiplicative group if $xy^{-1} \epsilon M$ whenever both x and y are in M.

It is clear that $\mathbb{C} \setminus \{0\}$, $\mathbb{R} \setminus \{0\}$, and $Q \setminus \{0\}$ are multiplicative groups. So are the sets $\{1\}$, $\{1, -1\}$, and T (the unit circle). We have an analogue of Lemma 1.

Lemma 2. Let M be a multiplicative group. Then (i) $1 \epsilon M$; (ii) for any $x \epsilon M$ the number x^{-1} is in M; (iii) when x and y are in M so is xy; (iv) for any $x \epsilon M$ the set $\{x^n \mid n \epsilon Z\}$ is contained in M.

When M, N are multiplicative groups and $N \subseteq M$, we shall say that N is a subgroup of M. In particular, when $x \epsilon M$ the set $\{x^n \mid n \epsilon Z\}$, which is clearly a multiplicative group, is a subgroup of M.

It is now necessary to specify the arithmetic operation under which a given set is a group. Unless it is clear, from the context, what operation we mean, we shall denote an additive group by $(A, +)$ and a multiplicative group by $(M, \cdot)$.

The exponential function, $f(x) = e^x$, maps $\mathbb{R}$ onto the positive real numbers. The set of all positive real numbers, let us denote this by $\mathbb{R}_+$, is clearly a multiplicative group. Thus f maps $(\mathbb{R}, +)$ onto $(\mathbb{R}_+, \cdot)$ and, furthermore, $f(x+y) = f(x) \cdot f(y)$ for all x, y in $\mathbb{R}$. This particular function is, of course, both one-to-one and onto. Let us look at this kind of mapping in a slightly more general context.

Let $(A, +)$ be an additive group, let $(M, \cdot)$ be a multiplicative group, and let $f : A \to M$ be a function such that $f(a+b) = f(a) f(b)$ for all a, b in A.

Because $0 \notin M$ and $f(a) = f(a+0) = f(a) \cdot f(0)$, we see that $f(0) = 1$. But then $1 = f(0) = f(a + (-a)) = f(a) f(-a)$, giving us the fact that $f(-a) = f(a)^{-1}$. Using these observations we can show that the range of f is a subgroup of $(M, \cdot)$. To do this suppose x, y elements in the range of f are given. We must show that xy^{-1} is in the range of f. But we have a, b in A such that $f(a) = x$ and $f(b) = y$. Thus $(a-b) \in A$, because A is an additive group, and $f(a-b) = f[a + (-b)] = f(a) f(-b) = xy^{-1}$, proving our claim.

Let us now show that $\{a \in A \mid f(a) = 1\}$ is a subgroup of $(A, +)$. Given a, b in this set (i.e., given a, b in A such that $f(a) = 1 = f(b)$) we must show that $a - b$ is in this set (i.e., we must show that $f(a-b) = 1$). But $f(a-b) = f[a + (-b)] = f(a) f(-b) = f(a) f(b)^{-1} = 1$ and we are finished.

Exercises 4

*1. Prove Lemma 2.

*2. The function $f(x) = \ln x$ maps $\mathbb{R}_+$ onto $\mathbb{R}$. Moreover, for any x, y in $\mathbb{R}_+$ we have $f(xy) = f(x) + f(y)$. This particular function is one-to-one and onto. Now let $(M, \cdot)$ be a multiplicative group, let $(A, +)$ be an additive group, and let $f: M \to A$ satisfy $f(xy) = f(x) + f(y)$ for all x, y in M.

   (a) Show that the range of f is a subgroup of $(A, +)$.
   (b) Show that $\{x \in M \mid f(x) = 0\}$ is a subgroup of $(M, \cdot)$.
   (c) Show that $U_4 = \{\pm 1, \pm i\}$ is a multiplicative group.
   (d) Let $f : \mathbb{Z} \to U_4$ be defined as follows: $f(n) = i^n$ for every n. Show that $f(m+n) = f(m) f(n)$ for all m, n in $\mathbb{Z}$. Identify the range of f and $\{n \in \mathbb{Z} \mid f(n) = 1\}$.
   (e) Let $g : \mathbb{Z} \to U_4$ be defined as follows: $g(n) = (-1)^n$ for every n. Show that $g(m+n) = g(m) g(n)$ for all m, n in $\mathbb{Z}$. Identify the range of g and $\{n \in \mathbb{Z} \mid g(n) = +1\}$.

*3. Let n be a fixed positive integer. Let $U_n = \{z \epsilon \mathbb{C} \mid z^n = 1\}$.

   (a) Show that U_n is a multiplicative group; in fact, it is a subgroup of T.

   (b) Let $p(x)$ be a given polynomial. The complex number r is a root of the equation $p(x) = 0$ if, and only if, $p(r) = 0$. We recall that r is a root of the equation $p(x) = 0$ if, and only if, $x - r$ is a factor of the polynomial $p(x)$. The largest positive integer k such that $(x - r)^k$ is a factor of $p(x)$ is called the multiplicity of the root r. A root of multiplicity one is called a simple root. Show that any root of $p(x) = 0$ that has multiplicity greater than one is also a root of $p'(x) = 0$ (the derivative of $p(x)$). Conclude that every root of $x^n - 1 = 0$ is simple and hence that U_n contains n distinct numbers.

4. Let $Q(\sqrt{2}) = \{a + b\sqrt{2} \mid a, b \text{ are in } Q\}$. Clearly, $Q(\sqrt{2}) \subseteq \mathbb{R}$.

   (a) Show that $Q(\sqrt{2})$ is a subgroup of $(\mathbb{R}, +)$.

   (b) Show that $Q(\sqrt{2}) \setminus \{0\}$ is a subgroup of $(\mathbb{R} \setminus \{0\}, \cdot)$.

*5. When $(M, \cdot)$ is a multiplicative group and $x \epsilon M$, the set $\{x^n \mid n \epsilon Z\} \subseteq M$, as we have seen. If there is an $x \epsilon M$ such that $\{x^n \mid n \epsilon Z\} = M$, then we say that $(M, \cdot)$ is a cyclic group and we say that x is a generator of M.

   (a) Show that $(U_4, \cdot)$ is cyclic and find all of its generators.

   (b) Show that $(U_n, \cdot)$ is a cyclic group and identify a generator. Hint: Every element of U_n is an element of T and hence can be written as $e^{i\theta}$.

   (c) An additive group $(A, +)$ is said to be cyclic if there is an $a \epsilon A$ such that $A = \{na \mid n \epsilon Z\}$. We have seen that any subgroup of Z is cyclic. Is Z cyclic?

5. Not So Naive Group Theory

We have seen that for any $f : \mathbb{R} \to \mathbb{K}$, the set $P(f) = \{\alpha \epsilon \mathbb{R} \mid f(x + \alpha) = f(x)$ for all $x\}$ is an additive group. Observe that for real numbers x, y we have $f(x) = f(y)$ when $x - y$ is in $P(f)$; for then $x - y = \alpha \epsilon P(f)$ and $f(x) = f(y + \alpha) = f(y)$. Thus f is constant on certain subsets of $\mathbb{R}$. We can get a better understanding of the structure of these sets by making use of the group theory we discussed in the last section.

Definition 1. Let A be an additive group and let C be a subgroup of A. We shall say that the elements a, b of A are congruent modulo C, and we shall write $a \equiv b \bmod C$, if $(a - b) \epsilon C$.

We shall give many examples of this very soon. For now we observe that congruence modulo C gives us a relation on the set A; it is in fact, as we shall now show, an equivalence relation (Chapter Zero, Section 2).

Lemma 1. For any additive group A and any subgroup C of A, congruence modulo C is an equivalence relation on A.

Proof. We must show that this relation is reflexive, symmetric, and transitive. To prove that it is reflexive we must show that for any $a \epsilon A$, $a \equiv a \mod C$; i.e., we must show that for any $a \epsilon A, (a-a)$ is in C. But this is immediate because $a - a = 0$ and 0 is in any subgroup of A (Section 4, Lemma 1, (i)).

Next we must show that for any $a, b \in A$ such that $a \equiv b \mod C$, we also have $b \equiv a \mod C$. Again this is immediate because $(a - b) \in C$ and so $-(a - b) = b - a$ is in C by Lemma 1, (ii), of Section 4.

Finally, suppose that a, b, c are in A and $a \equiv b \mod C$, $b \equiv c \mod C$. We must show that $a \equiv c \mod C$. The first of these gives us $(a - b) \in C$, and the second gives $(b - c) \in C$. By Lemma 1, (iii), of Section 4, we may add two elements of C. Thus $(a - b) + (b - c) = (a - c) \in C$. This, of course, says $a \equiv c \mod C$.

For any $a \epsilon A$, $[a] = \{b \epsilon A \mid a \equiv b \mod C\}$ is called the equivalence class containing a (Chapter Zero, Exercises 2, problem 1). The set of equivalence classes is denoted by A/C; read "$A \mod C$."

We began our discussion by noting that for any $f : \mathbb{R} \to \mathbb{K}$, the set $P(f)$ is a subgroup of the additive group $\mathbb{R}$ and that $f(x) = f(y)$ whenever $x \equiv y \mod P(f)$. Thus f is constant on the equivalence classes of $\mathbb{R}$; i.e., f is defined on the elements of $\mathbb{R}/P(f)$.

As an illustration of this, suppose f has smallest positive period 2π. Because $P(f)$ is then $\{(2\pi)n \mid n \epsilon Z\}$, we see that $x \equiv y \mod P(f)$ simply means $x - y = (2\pi)k$ for some integer k. But any real number x can be written as $x = (2\pi)k + r$ where k is an integer and $0 \leq r < 2\pi$. Thus each equivalence class contains a number (and clearly only one such number) from the set $[0, 2\pi)$. So we may regard f as being defined on this last set. This is, of course, just another way of saying that f is a function on the unit circle.

For subgroups of the additive group Z the construction of the equivalence classes is simple and illuminating. To be specific, we take $A = Z$ and $C = \{6n \mid n \epsilon Z\}$. Then $a \equiv b \mod C$ means $a - b$ is a multiple of six. Now for any $a \epsilon Z$ we may write $a = 6n + r$, where n is an integer, r is an integer, and $0 \leq r < 6$. Thus any $a \epsilon Z$ is equivalent to one, and clearly only one, of the numbers $0, 1, 2, 3, 4, 5$. Thus $A/C = \{[0], [1], [2], [3], [4], [5]\}$, where $[0] = \{n \epsilon Z \mid n \equiv 0 \mod C\}$....

The choice of six is simply one of convenience. We should note here that $C = \{6n \mid n \epsilon Z\}$ is usually denoted by (6) and the set $A/C = Z/(6)$ is often denoted by Z_6. Furthermore, $a \equiv b \mod C$ is often written $a \equiv b \mod 6$.

Given any subgroup C of the additive group Z, we have seen that $C = \{n\ell \mid n\epsilon Z\}$ where ℓ is the smallest positive member of C (Corollary 2 to Lemma 1 of Section 4). We shall denote this subgroup by (ℓ). Note that $Z/C = Z/(\ell)$ consists of the ℓ equivalence classes $[0], [1], ..., [\ell-1]$. This set is denoted by Z_ℓ.

Returning now to the case of an additive group A and a subgroup C of A, let us note that there is a natural "addition" on the elements of A/C. Given $[a], [b]$ in this set we choose an element of $[a]$, let us take a, and an element of $[b]$, take b, form their sum in A and set $[a] + [b]$ equal to the equivalence class containing $a + b$. To show that this is well defined, we must show that if $a'\epsilon(a)$ and $b'\epsilon(b)$, then $a' + b'$ and $a + b$ are in the same equivalence class; i.e., we must show that $(a+b) \equiv (a'+b') \mod C$. This is very easy. Because $a'\epsilon[a]$ and $b'\epsilon[b]$ we have $a - a'$ in C and $b - b'$ in C. Thus $(a - a') + (b - b')$ is in C (Section 4, Lemma 1, (iii)), which says $(a+b) \equiv (a'+b') \mod C$.

Let us illustrate this operation for the case of $Z_6 = Z/(6)$. To add $[3]$ and $[4]$ we take any element of $[3]$, let's take 3 (you can take 45 if you prefer), and any element of $[4]$, we'll take 4 but you can take 52 if you wish, and add them. So, because $3 + 4 = 7$, $[3] + [4] = [7] = [1]$ because $7 \equiv 1 \mod 6$ (note that $45 + 52 = 97$ so $[3] + [4] = [97]$, but $97 \equiv 1 \mod 6$).

Following this pattern we can construct an "addition table" for Z_6. It is customary, and convenient, to denote the elements of this set by $0, 1, 2, 3, 4, 5$. We have:

+	0	1	2	3	4	5
0	0	1	2	3	4	5
1	1	2	3	4	5	0
2	2	3	4	5	0	1
3	3	4	5	0	1	2
4	4	5	0	1	2	3
5	5	0	1	2	3	4

We should stress that the elements of Z_6 are *not* complex numbers and the operation $+$ is *not* ordinary addition. The table shows that 0 is the identity element of Z_6. We mean $a + 0 = 0 + a = a$ for every $a\epsilon Z_6$. Furthermore, given any $a\epsilon Z_6$ there is a $b\epsilon Z_6$ such that $a + b = 0$.

We call b the additive inverse of a. So we can define a kind of subtraction: $a - b$ is taken to mean a plus the additive inverse of b. For example, $3 - 2$ is to mean $3 + 4$ because 4 is the additive inverse of 2. Thus $3 - 2$ is 1 while $3 - 4 = 3 + 2 = 5$.

There is a natural map $\varphi_6 : Z \to Z_6$. For any $n\epsilon Z$ we take $\varphi_6(n)$ to be the remainder one gets when n is divided by 6. So $\varphi_6(12) = 0$, $\varphi_6(11) = 5$, $\varphi_6(7) = 1$, etc. Observe that, for any m, n in Z, $\varphi_6(m+n) = \varphi_6(m) + \varphi_6(n)$.

We warn the reader that in writing $m+n$ we are using ordinary addition, whereas in writing $\varphi_6(m) + \varphi_6(n)$ we are using the operation defined in Z_6. Interpreting this sum as an ordinary addition leads to erroneous results.

Exercises 5

*1. Let M be a multiplicative group and let N be a subgroup of M.

   (a) For any s, t in M define $s \equiv t \bmod N$ to mean $st^{-1} \epsilon N$. Show that congruence modulo N is an equivalence relation on M.

   (b) The set of equivalence classes of M under the relation defined in (a) is denoted by M/N. For any $[s], [t]$ in M/N define $[s] * [t]$ as follows: choose $s\epsilon[s]$, $t\epsilon[t]$ and set $[s] * [t] = [st]$. Show that this operation is well defined; i.e., if $s' \epsilon [s]$ and $t' \epsilon [t]$ show that $s't' \equiv st \bmod N$.

   (c) Show that the element $[1] \epsilon M/N$ is a multiplicative identity for this set; i.e., show that $[s] * [1] = [1] * [s] = [s]$ for all $[s]$ in M/N.

   (d) For each $[s] \epsilon M/N$ show that there is a $[t] \epsilon M/N$ such that $[s] * [t] = [1]$.

   (e) Let M be the multiplicative group of all nonzero complex numbers and let $N = T$ (the unit circle). Identify the elements of M/N.

2. Compute the addition table for Z_9. Define $\varphi_9(n)$ to be the remainder one gets when n is divided by 9. Show that when $n > 0$, $\varphi_9(n)$ can be computed as follows: Add the digits of n. If the result is greater than 9, add its digits. Continue until you get a number between 0 and 9.

3. (a) We refer to problem 2. Show that for all m, n in Z, $\varphi_9(m+n) = \varphi_9(m) + \varphi_9(n)$.

   (b) Compute $\varphi_9(2315) + \varphi_9(7489)$. Show that this is $\varphi_9(9804)$.

   (c) Find $\{n \epsilon Z \mid \varphi_9(n) = 0\}$.

*4. Let A be an additive group and let C be a subgroup of A. Define $\varphi : A \to A/C$ as follows: $\varphi(a) = [a]$ for every $a \epsilon A$.

   (a) Show that $\varphi(a+b) = \varphi(a) + \varphi(b)$ for all $a, b \epsilon A$.

   (b) What is $\{a \epsilon A \mid \varphi(a) = [0]\}$?

6. Finite Fourier Transform

The Fourier transform plays a fundamental role in modern digital processing (see Notes). This application requires the handling of enormous amounts of numerical data. I have been told that a color TV picture, for example, requires the processing of about 8 million bits of data per second. What is clearly needed here is an efficient numerical technique for computing the Fourier transforms involved because, in most cases, they cannot be found analytically. The fast Fourier transform algorithm is such a technique. Its efficacy is remarkable. In some cases 1 billion computations can be replaced by 1 million, one one-thousandth of the original number. What one does is replace the Fourier transform of a function defined on $\mathbb{R}$ by the Fourier transform of a function defined on Z_ℓ for large ℓ. The algorithms are then applied to this replacement. We shall concern ourselves with the problem of how the Fourier transform can be defined on Z_ℓ. Once we have done that, we investigate its properties and give some applications.

In the case of the additive group Z we considered

$$\ell^1(Z) = \left\{ a : Z \to \mathbb{C} \mid \sum_{-\infty}^{\infty} \mid a(n) \mid < \infty \right\}$$

For a, b in $\ell^1(Z)$ we defined $(a * b)(n) = \sum_{-\infty}^{\infty} a(n-m) b(m)$ and note we need here the group operation in Z; i.e., $n - m = n + (-m)$ where $(-m)$ is the additive inverse of m. We have seen that subtraction is possible in Z_ℓ and so we can define convolution for functions on this set. Let us do that and work out a simple case in detail. To begin with let

$$\ell^1(Z_\ell) = \left\{ a : Z_\ell \to \mathbb{C} \mid \sum_{n=0}^{\ell-1} \mid a(n) \mid < \infty \right\}$$

Clearly, every function on Z_ℓ is in $\ell^1(Z_\ell)$, so our notation is a little silly. We use it to emphasize the analogy between what we are doing here and what we did on Z.

Definition 1. For any $a, b \epsilon \ell^1(Z_\ell)$ we define the convolution of a and b, $a * b$, as follows: $(a * b)(k) = \sum_{m=0}^{\ell-1} a(k-m) b(m)$.

Let us stress that the subtraction is to take place in Z_ℓ. To get a "feeling" for this operation, let us work out the convolution of two elements a, b in $\ell^1(Z_3)$.

We have:

$$(a*b)(0) = \sum_{j=0}^{2} a(0-j)b(j) = a(0-0)b(0) + a(0-1)b(1) + a(0-2)b(2)$$
$$= a(0)b(0) + a(2)b(1) + a(1)b(2)$$

because $0-1$ is the additive inverse of 1 in Z_3 and this is 2, while the additive inverse of 2 is 1. Similarly,

$$(a*b)(1) = \sum_{j=0}^{2} a(1-j)b(j) = a(1-0)b(0) + a(1-1)b(1) + a(1-2)b(2)$$
$$= a(1)b(0) + a(0)b(1) + a(2)b(2)$$

because, in Z_3, $1-2 = 1+1 = 2$.

$$(a*b)(2) = \sum_{j=0}^{2} a(2-j)b(j) = a(2-0)b(0) + a(2-1)b(1) + a(2-2)b(2)$$
$$= a(2)b(0) + a(1)b(1) + a(0)b(2)$$

Observe that if we write a and b as (a_0, a_1, a_2) and (b_0, b_1, b_2), respectively, then $a*b - (a_0 b_0 + a_2 b_1 + a_1 b_2, a_1 b_0 + a_0 b_1 + a_2 b_2, a_2 b_0 + a_1 b_1 + a_0 b_2)$.

This last expression clearly shows that $\ell^1(Z_3)$ contains an identity. The function $e = (1, 0, 0)$ satisfies $a*e = e*a = a$ for every a.

This operation is sometimes referred to as "cyclic convolution." Its basic properties are easily established.

Lemma 1. For any a, b, c in $\ell^1(Z_\ell)$ we have

(a) $a*b = b*a$;

(b) $a*(b*c) = (a*b)*c$;

(c) $\alpha(a*b) = (\alpha a)*b = a*(\alpha b)$ for any $\alpha \epsilon \mathbb{C}$;

(d) $a*(b+c) = a*b + a*c$;

(e) There is an element $e \epsilon \ell^1(Z_\ell)$ such that $a*e = e*a = a$ for every a.

The Fourier transform on Z takes each function in $\ell^1(Z)$ to a continuous function on T; the unit circle in the complex plane. Recall that for $a \epsilon \ell^1(Z)$, $\hat{a}(\theta) = \sum_{-\infty}^{\infty} a(n) e^{in\theta}$. Let us examine, a little more closely, $e^{in\theta}$ for θ fixed. We can define $c_\theta : Z \to T$ by setting $c_\theta(n) = e^{in\theta}$ for each $n \epsilon Z$; remember θ is fixed. Observe that $c_\theta(m+n) = e^{i(m+n)\theta} = e^{im\theta} \cdot e^{in\theta} = c_\theta(m) \cdot c_\theta(n)$ for all m, n in Z. Maps like this have a name.

Definition 2. A map $c : Z \to T$ such that $c(m+n) = c(m) \cdot c(n)$ for all $m, n \epsilon Z$ is called a character of Z.

Lemma 2. To every character c of Z there corresponds a unique real number $\theta \epsilon [0, 2\pi)$ such that $c(n) = \left(e^{i\theta}\right)^n$ for every $n \epsilon Z$.

Proof. Because $c(n)$ is never zero and $c(n+0) = c(n) \cdot c(0)$, we see that $c(0) = 1$. Furthermore, $c(n) = c(1+1+\ldots+1) = c(1)^n$ for every natural number n. Combining these observations we have $1 = c(0) = c(n+(-n)) = c(n) \cdot c(-n)$ and so $c(n) = c(1)^n$ for every integer n.

Now $c(1) \in T$ and so $c(1) = e^{iy}$ for some real number y. If $y \notin [0, 2\pi)$ we may write $y = (2\pi)k + \theta$ where k is an integer and $\theta \in [0, 2\pi)$. Clearly, then $c(n) = \left(e^{iy}\right)^n = e^{i(2k\pi+\theta)n} = e^{i2kn\pi}\left(e^{i\theta}\right)^n = \left(e^{i\theta}\right)^n$ for every $n \in Z$. This proves the lemma because it is clear that no two θ's in $[0, 2\pi)$ can give the same character of Z.

Corollary 1. There is a one-to-one correspondence between the characters of Z and the points of T.

Proof. Because there is a one-to-one correspondence between the points of T and the set $[0, 2\pi)$, this follows immediately from our lemma.

For $a \epsilon \ell^1(Z)$, $\hat{a}(\theta) = \sum_{-\infty}^{\infty} a(n) e^{in\theta}$. If we think of θ as fixed, then we are taking the character $e^{i\theta}$ and applying it to each n, $c_\theta(n) = e^{in\theta}$, then multiplying by $a(n)$ and finally summing over the domain of a; i.e., summing over Z. We will do something similar for functions in $\ell^1(Z_\ell)$. To begin, however, we must first identify the "characters of Z_ℓ."

Definition 3. Let ℓ be a natural number and consider the set Z_ℓ. A map $\gamma : Z_\ell \to T$ is called a character of Z_ℓ if $\gamma(m+n) = \gamma(m) \cdot \gamma(n)$ for all m, n in Z_ℓ.

Lemma 3. There is a one-to-one correspondence between the characters of Z_ℓ and the solutions to the equation $z^\ell = 1$.

Proof. Let w be a complex number such that $w^\ell = 1$. Define a map $\gamma : Z_\ell \to T$ by setting $\gamma(k) = w^k$ for each $k \epsilon Z_\ell$. Clearly, $\gamma(m+n) = w^{m+n} = w^m w^n = \gamma(m)\gamma(n)$ for any m, n in Z_ℓ. Furthermore, if $m \equiv n \bmod \ell$, then $\gamma(m) = \gamma(n)$, so γ is well defined and is clearly a character of Z.

Suppose now that γ is a given character of Z_ℓ. Then because $\gamma(k+0) = \gamma(k)\gamma(0)$ we see that $\gamma(0) = 1$. Also, for any $k \epsilon Z_\ell$ we have $\gamma(k) = \gamma(1+\ldots+1) = \gamma(1)^k$. Thus γ is completely determined by $\gamma(1)$. But $\gamma(\ell) = \gamma(0) = \gamma(1)^\ell = 1$ and so $\gamma(1)$ is a root of the equation $z^\ell = 1$.

Definition 4. Let ℓ be a natural number. The set of all solutions to the equation $z^\ell = 1$ is denoted by U_ℓ. The elements of this set are called ℓth roots of unity.

We have seen (Exercises 4, problem 3) that U_ℓ is a multiplicative group containing ℓ elements.

Let us look more closely at U_4 and its multiplication table:

$\cdot$	1	-1	i	$-i$
1	1	-1	i	$-i$
-1	-1	1	$-i$	i
i	i	$-i$	-1	1
$-i$	$-i$	i	1	-1

It is, of course, true that $1^4 = 1$, but 4 is *not* the smallest positive integer for which this is true. Clearly, $1^1 = 1$; we say 1 has order 1. Similarly, $(-1)^4 = 1$, but the smallest positive integer for which this is true is 2; we say (-1) has order 2. The remaining elements i and $-i$ each has order 4; 4 is the smallest positive integer n such that $i^n = 1$ and it is the smallest positive integer n each that $(-i)^n = 1$.

Note that $\{i^n \mid n \epsilon Z\}$ is just $\{1, -1, i, -i\}$; i.e., it coincides with U_4. Thus U_4 is a cyclic group and i is a generator of this group (Exercises 4, problem 5). Clearly, $-i$ is also a generator of this group.

Definition 5. Let ℓ be a natural number and let $\alpha \epsilon U_\ell$. The order of α, $\theta(\alpha)$, is the smallest natural number k such that $\alpha^k = 1$. When $\theta(\alpha) = \ell$ we say that α is a primitive ℓth root of unity.

Clearly, $e^{2\pi i/\ell} = \cos\frac{2\pi}{\ell} + i\sin\frac{2\pi}{\ell}$ is a primitive ℓth root. When $\ell = 3$, this gives us $\cos\frac{2\pi}{3} + i\sin\frac{2\pi}{3} = -\frac{1}{2} + i\frac{\sqrt{3}}{2}$. When $\ell = 4$, we get $\cos\frac{2\pi}{4} + i\sin\frac{2\pi}{4} = i$.

Lemma 4. The order of each element of U_ℓ is a divisor of ℓ. An element w of U_ℓ is a primitive ℓth root if, and only if, it generates the group U_ℓ.

Proof. If $\alpha \epsilon U_\ell$ and $\theta(\alpha) = k$, then clearly $k \leq \ell$. Now $\ell = qk + r$ where q and r are integers and $0 \leq r < k$. Thus $1 = \alpha^\ell = \alpha^{qk+r} = (\alpha^k)^q \alpha^r = (1)^q \alpha^r = \alpha^r$, which says, because k is the smallest natural number such that $\alpha^k = 1$, that $r = 0$.

Now suppose that $w \epsilon U_\ell$ has order ℓ. Then $U_\ell \supseteq \{w^i \mid 1 \leq j \leq \ell\}$ so to show that these sets are equal we need only show that the last one contains ℓ distinct members. But if $w^p = w^q$ for $1 \leq p < q \leq \ell$, then $w^{q-p} = 1$ and $0 < q - p < \ell$. This contradicts the fact that w has order ℓ. Thus w generates U_ℓ.

Finally, let w be a generator of U_ℓ. If $\theta(w) = k < \ell$, then $\{w^i \mid j \epsilon Z\} = \{w^1, w^2, w^3, ..., w^k\}$, and this contains $k < \ell$ elements. So the order of w must be ℓ.

Definition 6. Let ℓ be a fixed natural number and let w be the primitive ℓth root $e^{2\pi i/\ell}$. For any $a \epsilon \ell^1(Z_\ell)$ we define $\hat{a}(w^k) = \sum_{j=0}^{\ell-1} a(j)(w^k)^j$ and call this function, its domain is U_ℓ, the Fourier transform of a.

For the case of a function $a \epsilon \ell^1(Z_3)$ this becomes

$$\hat{a}(w^0) = \sum_{j=0}^{2} a(j)(w^0)^j = a(0) + a(1) + a(2)$$

$$\hat{a}(w) = \sum_{j=0}^{2} a(j) w^j = a(0) + a(1)w + a(2)w^2$$

$$\hat{a}(w^2) = \sum_{j=0}^{2} a(j)(w^2)^j = a(0) + a(1)w^2 + a(2)w^4$$

A more convenient way to write this is

$$\begin{pmatrix} \hat{a}(w^0) \\ \hat{a}(w) \\ \hat{a}(w^2) \end{pmatrix} = \begin{pmatrix} 1 & 1 & 1 \\ 1 & w & w^2 \\ 1 & w^2 & w^4 \end{pmatrix} \begin{pmatrix} a(0) \\ a(1) \\ a(2) \end{pmatrix}$$

Of course, because $w = e^{2\pi i/3} \epsilon U_3$, $w^4 = w$.

Recall that the map $\mathcal{F}$ takes $\ell^1(Z)$ into $C(T)$. So if $a \epsilon \ell^1(Z)$, $\mathcal{F}(a) = \hat{a}$ is defined on T. Furthermore, $\mathcal{F}$ is linear, invertible, and "preserves" convolution. The map $\mathcal{F}_\ell$ takes each function $a \epsilon \ell^1(Z_\ell)$ and transforms it into a function $\hat{a}$ on U_ℓ. However, it is customary, or at least very common, to regard $\hat{a}$ as also having domain Z_ℓ. This is done as follows: Recall $\gamma(k) = w^k$ is a map from Z_ℓ to U_ℓ that satisfies $\gamma(k+m) = \gamma(k)\gamma(m)$ all k, m in Z_ℓ. When w is primitive, as it is in this discussion, γ is both one-to-one and onto. Thus we may identify $(Z_\ell, +)$ with $(U_\ell, \cdot)$. With this understanding $\hat{a}(w^0) = \hat{a}(0)$, $\hat{a}(w) = \hat{a}(1), ..., \hat{a}(w^{\ell-1}) = \hat{a}(\ell-1)$.

Our Fourier transform may now be written $\mathcal{F}_\ell(a)(k) = \hat{a}(k) = \sum_{j=0}^{\ell-1} a(j)(w^k)^j = \sum_{j=0}^{\ell-1} a(j)\left(e^{\frac{2\pi i}{\ell}}\right)^{kj}$ in matrix notation:

$$\begin{pmatrix} \hat{a}(0) \\ \hat{a}(1) \\ \vdots \\ \hat{a}(\ell-1) \end{pmatrix} = \begin{pmatrix} 1 & 1 & 1 & \cdots & 1 \\ 1 & w & w^2 & \cdots & w^{\ell-1} \\ \cdots & & & & \\ 1 & (w^{\ell-1}) & (w^{\ell-1})^2 & & (w^{\ell-1})^{\ell-1} \end{pmatrix} \begin{pmatrix} a(0) \\ a(1) \\ \vdots \\ a(\ell-1) \end{pmatrix}$$

Theorem 1. For each fixed ℓ the map $\mathcal{F}_\ell$ is linear and preserves convolution; i.e., $\mathcal{F}_\ell(a * b) = \mathcal{F}_\ell(a) \cdot \mathcal{F}_\ell(b)$.

Proof.

$$\mathcal{F}_\ell(a*b)(n) = \sum_{k=0}^{\ell-1} (a*b)(k)(w^n)^k = \sum_{k=0}^{\ell-1}\left[\sum_{j=0}^{\ell-1} a(j)b(k-j)\right](w^n)^k$$

$$= \sum_{j=0}^{\ell-1} a(j)\left[\sum_{k=0}^{\ell-1} b(k-j)(w^n)^k\right]$$

$$= \left[\sum_{j=0}^{\ell-1} a(j)(w^n)^j\right]\left[\sum_{k=0}^{\ell-1} b(k-j)(w^n)^{k-j}\right]$$

$$= \mathcal{F}_\ell(a)(n) \cdot \mathcal{F}_\ell(b)(n)$$

Let us set $\mathcal{F}_\ell^{-1}(a) = \frac{1}{\ell}\sum_{j=0}^{\ell-1} a(j)(\overline{w}^k)^j$. Then

Theorem 2. For any $a \in \ell^1(Z_\ell)$ we have $\hat{a}(k) = \mathcal{F}_\ell(a)(k) = \sum_{j=0}^{\ell-1} a(j)(w^k)^j$

and $a(k) = \mathcal{F}_\ell^{-1}(\hat{a})(k) = \frac{1}{\ell}\sum_{j=0}^{\ell-1} \hat{a}(j)(\overline{w}^k)^j$.

Proof. In this last sum we replace $\hat{a}(j)$ by its definition obtaining

$$\frac{1}{\ell}\sum_{j=0}^{\ell-1} \hat{a}(j)(\overline{w}^k)^j = \frac{1}{\ell}\sum_{j=0}^{\ell-1}\left(\sum_{n=0}^{\ell-1} a(n)(w^j)^n\right)(\overline{w}^k)^j$$

$$\frac{1}{\ell}\sum_{n=0}^{\ell-1} a(n)\left[\sum_{j=0}^{\ell-1}(w^{n-k})^j\right]$$

When $n = k$ this inner sum is ℓ, and when $n \neq k$ it is zero (Exercises 6, problem 1).
Thus

$$\frac{1}{\ell}\sum_{j=0}^{\ell-1} \hat{a}(j)(\overline{w}^k)^j = a(k)$$

We have seen that the Fourier transform on Z_ℓ, the finite Fourier transform, maps functions defined on Z_ℓ to functions defined on the set of characters of Z_ℓ, the set $\widehat{Z}_\ell$. Now $\widehat{Z}_\ell$ can be identified with U_ℓ, and this last set can be identified with Z_ℓ. So the transform $\Im_\ell$ maps functions defined on Z_ℓ to functions defined on Z_ℓ. The transform is linear and so in this, finite, case it can be represented by a matrix. Moreover, the transform preserves convolution and is invertible.

We shall see (Exercises 6, problem 1) that when $\omega \in U_\ell$ (i.e., when ω is an ℓth root of unity) and $\omega \neq 1$, then $\sum_{j=0}^{\ell-1} \omega^j = 0$. Now suppose that ω is a primitive ℓth root of unity. Then for any m, $0 < m \leq \ell - 1$, ω^m is in U_ℓ and, because ω is primitive, $\omega^m \neq 1$. Hence $\sum_{j=0}^{\ell-1} (\omega^m)^j = 0$.

Let us work out the case $\ell = 3$. Recall:

$$\Im_3(f) = \widehat{f} = \begin{pmatrix} 1 & 1 & 1 \\ 1 & \omega & \omega^2 \\ 1 & \omega^2 & \omega^4 \end{pmatrix} f$$

where, of course, $\omega^4 = \omega$. Consider now

$$\begin{pmatrix} 1 & 1 & 1 \\ 1 & \omega & \omega^2 \\ 1 & \omega^2 & \omega^4 \end{pmatrix} \begin{pmatrix} 1 & 1 & 1 \\ 1 & \bar{\omega} & \bar{\omega}^2 \\ 1 & \bar{\omega}^2 & \bar{\omega}^4 \end{pmatrix} = \begin{pmatrix} 3 & 1+\bar{\omega}+\bar{\omega}^2 & 1+\bar{\omega}^2+\bar{\omega}^4 \\ 1+\omega+\omega^2 & 3 & 1+\bar{\omega}+\bar{\omega}^2 \\ 1+\omega^2+\omega^4 & 1+\omega+\omega^2 & 3 \end{pmatrix}$$

we have used the fact that $\bar{\omega} = 1/\omega$ (Exercises 6, problem 2(b)).
This last matrix is just

$$3 \begin{pmatrix} 1 & 0 & 0 \\ 0 & 1 & 0 \\ 0 & 0 & 1 \end{pmatrix}$$

and, it is easy to see, reversing the factors gives us the same result. Thus to invert $\Im_3$ we take its matrix and replace each entry by its complex conjugate and multiply the result by $\frac{1}{3}$. In another notation:

$$\Im_3(f)(\bar{k}) = \widehat{f}(\bar{k}) = \sum_{j=0}^{2} f(\bar{j})(\omega^k)^j = \sum_{j=0}^{2} f(\bar{j})(e^{+\frac{2\pi i}{3}})^{kj}$$

and

$$\mathfrak{F}_3^{-1}(g)(\bar{k}) = \frac{1}{3}\sum_{j=0}^{2} g(\bar{j})(\bar{\omega}^k)^j = \frac{1}{3}\sum_{j=0}^{2} g(\bar{j})(e^{-\frac{2\pi i}{3}})^{kj}$$

Remark on Notation. Some writers define the Fourier transform in a slightly different way to obtain a symmetry between the transform and its inverse. They set

$$\mathfrak{F}_\ell(f) = \widehat{f}(k) = \frac{1}{\sqrt{\ell}} \sum_{j=0}^{\ell-1} \widehat{f}(j)(\omega^k)^j$$

and obtain

$$\mathfrak{F}_\ell^{-1}(g) = \frac{1}{\sqrt{\ell}} \sum_{j=0}^{\ell-1} g(j)(\bar{\omega}^k)^j$$

This is also done in connection with the Fourier transform on $\mathbb{R}$. One writes

$$\mathfrak{F}(f) = \widehat{f}(x) = \frac{1}{\sqrt{2\pi}} \int_{-\infty}^{\infty} f(t) e^{+ixt} dt$$

so that

$$\mathfrak{F}^{-1}(g) = \frac{1}{\sqrt{2\pi}} \int_{-\infty}^{\infty} g(x) e^{-ixt} dx$$

We see little advantage in doing this, but the reader should be aware that it is not uncommon.

Exercises 6

*1. Let $\omega \in U_\ell$, $\omega \neq 1$. One of the properties of ℓth roots of unity that helps make the fast Fourier transform algorithm so efficient is the following: $\sum_{j=0}^{\ell-1} \omega^j = 0$. Follow the steps below to give two proofs of this.

  (a) Set $r = \sum_{j=0}^{\ell-1} \omega^j$. Show that $\omega r = r$ and hence, if $r \neq 0$, we have a contradiction.
  (b) Factor $\omega^n - 1$.

*2. Let $\omega \in U_\ell$. Show that (a) $\omega^{-1} = \omega^{\ell-1}$; (b) $\bar{\omega} = 1/\omega$; (c) if ℓ is a prime (an integer larger than one that has no factors other than itself and one), then every element of U_ℓ except $+1$ is a primitive ℓth root of unity.

3. Show that $\ell^1(Z_\ell)$ contains an element e such that $e * f = f$ for every $f \in \ell^1(Z_\ell)$. Compute $\Im_\ell(e)$. An element f of $\ell^1(Z_\ell)$ is said to be invertible if there is an element $g \in \ell^1(Z_\ell)$ such that $f * g = e$.

   (a) Show that f is invertible if, and only if, $\Im_\ell(f)$ is never zero.

   (b) If f is *not* invertible find all g such that $f * g = 0$.

   (c) f is called a divisor of zero if there is $g \neq 0$ such that $f * g = 0$. Show that f is a divisor of zero if, and only if, f is not invertible.

   (d) If f is invertible we have an element $f^{-1} \in \ell^1(Z_\ell)$ such that $f * f^{-1} = e$. Show that f^{-1} is unique and that $\Im_\ell(f^{-1})(\bar{n}) = 1/\Im_\ell(f)(\bar{n})$ for every $\bar{n} \in Z_\ell$.

4. Write the Fourier transform on $\ell^1(Z_2)$ in matrix form and then invert the matrix.

5. (a) Let $f \in \ell^1(Z_2)$ be given by $f = (0, 1)$. Compute $\Im_2(f)$. Similarly, compute $\Im_3(f)$ where $f = (0, 0, 1)$.

   (b) Characterize all $g \in \ell^1(Z_\ell)$ such that $g * g * \cdots * g$ (ℓ factors) is e, in terms of $\Im_\ell(g)$.

6. Given $f = (a_0, a_1, a_2)$, $g = (b_0, b_1, b_2)$ in $\ell^1(Z_3)$ compute:

   (a) $\hat{f}$ and $\hat{g}$;

   (b) $\hat{f} \cdot \hat{g}$;

   (c) $f * g$;

   (d) $\widehat{(f * g)}$.

7. Continuous characters of $\mathbb{R}$.
   Let $c : \mathbb{R} \to T$ be a continuous mapping such that $c(x+y) = c(x) \cdot c(y)$ for all x, y in R. Show that there is an element $\alpha \in \mathbb{R}$ such that $c(x) = \left(e^{i\alpha}\right)^x$ for all $x \in \mathbb{R}$. Hint: Recall that Q is dense in $\mathbb{R}$ (Chapter Zero, Section 3, Corollary 2 to Theorem 1).

7. An Application

There are some interesting, and surprising, connections between the finite Fourier transform and the theory of polynomial equations. These were found by Lagrange in his study of the cubic and quartic. The calculations involved are a bit messy, and we shall try to present the underlying ideas first so as not to lose sight of them in the technical details. Let us start with the simplest case of all, the well-known quadratic.

Suppose we are given the equation
$$ax^2 + bx + c = 0 \tag{1}$$
where $a \neq 0$ and the roots are x_1, x_2. The French mathematician Vieta first noted the relation between the roots and coefficients. If this case they are
$$x_1 + x_2 = \frac{-b}{a} \tag{2}$$
$$x_1 x_2 = \frac{c}{a} \tag{3}$$

As functions of x_1, x_2 these are clearly symmetric, meaning they are invariant when we interchange the two variables. It turns out that any symmetric function of x_1, x_2 can be represented in terms of the two functions given in (2) and (3); i.e., any symmetric function of the roots can be expressed in terms of the coefficients. We shall prove this only for the special cases we need for our discussion. As an example, consider $\Delta = (x_1 - x_2)^2$. This is clearly invariant when we reverse our variables. We can write Δ in terms of a, b, and c as follows:

$$\Delta = (x_1 - x_2)^2 = x_1^2 + x_2^2 - 2x_1 x_2 = x_1^2 + x_2^2 + 2x_1 x_2 - 4x_1 x_2 \tag{4}$$
$$= (x_1 + x_2)^2 - 4x_1 x_2$$
$$= \left(\frac{-b}{a}\right)^2 - 4\frac{c}{a} = \frac{b^2 - 4ac}{a^2}$$

where we have used (2) and (3). Combining (2) and (4) we have the system
$$x_1 + x_2 = -\frac{b}{a} \tag{5}$$
$$x_1 - x_2 = \frac{\sqrt{b^2 - 4ac}}{a}$$

This is easily solved, but note that it is equivalent to:
$$\begin{pmatrix} 1 & 1 \\ 1 & -1 \end{pmatrix} \begin{pmatrix} x_1 \\ x_2 \end{pmatrix} = \begin{pmatrix} -\frac{b}{a} \\ \frac{\sqrt{b^2 - 4ac}}{a} \end{pmatrix} \tag{6}$$

The coefficient matrix in (6) is the finite Fourier transform obtained using the primitive square root of unity. We may solve (1) by solving (5) or (6). In the latter case
$$\begin{pmatrix} x_1 \\ x_2 \end{pmatrix} = \frac{1}{2} \begin{pmatrix} 1 & 1 \\ 1 & -1 \end{pmatrix} \begin{pmatrix} -\frac{b}{a} \\ \frac{\sqrt{b^2 - 4ac}}{a} \end{pmatrix} \tag{7}$$

7. An Application

Let us turn now to the cubic

$$x^3 + px^2 + qx + r = 0 \tag{8}$$

with roots x_1, x_2, x_3. Lagrange chose, and fixed, a primitive cube root of unity w and considered the form

$$\Delta_1 = (x_1 + wx_2 + w^2 x_3)^3 \tag{9}$$

He noted that this is *not* symmetric. Three of the six permutations of x_1, x_2, x_3 do leave it invariant, but three transform it into

$$\Delta_2 = (x_1 + w^2 x_2 + w x_3)^3 \tag{10}$$

Similarly, the three permutations that leave (9) invariant also leave (10) invariant, whereas the other three transform it into (9). This means that $\Delta_1 + \Delta_2$ and $\Delta_1 \Delta_2$ are invariant under all six permutations and hence can be expressed in terms of p, q, and r; we shall do this explicitly below. It follows then that Δ_1 and Δ_2 can be found by solving the equation

$$\overline{X}^2 - (\Delta_1 + \Delta_2)\overline{X} + (\Delta_1 \Delta_2) = 0 \tag{11}$$

and once we have those numbers we may write

$$\begin{aligned} x_1 + x_2 + x_3 &= -p \\ x_1 + wx_2 + w^2 x_3 &= \sqrt[3]{\Delta_1} \\ x_1 + w^2 x_2 + wx_3 &= \sqrt[3]{\Delta_2} \end{aligned} \tag{12}$$

or

$$\begin{pmatrix} 1 & 1 & 1 \\ 1 & w & w^2 \\ 1 & w^2 & w \end{pmatrix} \begin{pmatrix} x_1 \\ x_2 \\ x_3 \end{pmatrix} = \begin{pmatrix} -p \\ \sqrt[3]{\Delta_1} \\ \sqrt[3]{\Delta_2} \end{pmatrix} \tag{13}$$

Again we see that the coefficient matrix is the finite Fourier transform. As we have seen in Section 6, the system (13) can be solved by inverting the matrix. We find that the solutions to (8) are

$$\begin{pmatrix} x_1 \\ x_2 \\ x_3 \end{pmatrix} = \frac{1}{3} \begin{pmatrix} 1 & 1 & 1 \\ 1 & \overline{w} & \overline{w}^2 \\ 1 & \overline{w}^2 & \overline{w} \end{pmatrix} \begin{pmatrix} -p \\ \sqrt[3]{\Delta_1} \\ \sqrt[3]{\Delta_2} \end{pmatrix} \tag{14}$$

We now show how one can express Δ_1 and Δ_2 in terms of p, q, and r. Recall x_1, x_2, x_3 are roots of

$$x^3 + px^2 + qx + r = 0 \tag{8}$$

The Vieta formulas in this case are

$$x_1 + x_2 + x_3 = -p \tag{15}$$
$$x_1 x_2 + x_1 x_3 + x_2 x_3 = q \tag{16}$$
$$x_1 x_2 x_3 = -r \tag{17}$$

Observe that $(-p)^2 = (x_1+x_2+x_3)^2 = x_1^2+x_2^2+x_3^2+2(x_1x_2+x_1x_3+x_2x_3)$ and hence, using (16), we have

$$x_1^2 + x_2^2 + x_3^2 = p^2 - 2q \tag{18}$$

Now if we put x_1, x_2, x_3 into (8) and add the results we get

$$x_1^3 + x_2^3 + x_3^3 + p(x_1^2 + x_2^2 + x_3^2) + q(x_1 + x_2 + x_3) + 3r = 0$$

and putting (15) and (18) into this equation we find

$$x_1^3 + x_2^3 + x_3^3 = -p(p^2 - 2q) + pq - 3r = -p^3 + 3pq - 3r \tag{19}$$

We are trying to express $\Delta_1 \Delta_2$ and $\Delta_1 + \Delta_2$ in terms of p, q, and r. The results obtained above enable us to do this for $\Delta_1 \Delta_2$ rather easily. Let

$$\delta = (x_1 + wx_2 + w^2 x_3)(x_1 + w^2 x_2 + wx_3)$$
$$= x_1^2 + x_2^2 + x_3^2 + w(x_1 x_3 + x_1 x_2 + x_2 x_3) + w^2(x_1 x_3 + x_1 x_2 + x_2 x_3) \tag{20}$$
$$= (p^2 - 2q) + (w + w^2)q$$

where we have used (18) and (16). Now w is a primitive cube root of unity, hence not equal to 1. Thus $w^2 + w + 1 = 0$ (Exercises 6, problem 1). Using this in (20), we get

$$\delta = p^2 - 2q - q = p^2 - 3q$$

and so

$$\Delta_1 \Delta_2 = (x_1 + wx_2 + w^2 x_3)^3 (x_1 + w^2 x_2 + wx_3)^3 = (p^2 - 3q)^3 \tag{21}$$

The expression for $\Delta_1 + \Delta_2$ in terms of p, q, and r is more difficult to obtain. We need a preliminary result first. Let

$$\lambda = x_1^2 x_2 + x_1 x_2^2 + x_1^2 x_3 + x_1 x_3^2 + x_2^2 x_3 + x_2 x_3^2 \tag{22}$$

7. An Application

By equations (15) and (16) we may write

$$-pq = (x_1 + x_2 + x_3)(x_1x_2 + x_1x_3 + x_2x_3)$$
$$= x_1^2x_2 + x_1^2x_3 + x_1x_2x_3$$
$$+ x_1x_2^2 + x_1x_2x_3 + x_2^2x_3$$
$$+ x_1x_2x_3 + x_1x_3^2 + x_2x_3^2$$
$$= \lambda - 3r$$

Hence

$$\lambda = -pq + 3r \tag{23}$$

A direct calculation shows that

$$(a+b+c)^3 = a^3 + b^3 + c^3 + 3(a^2b + ab^2 + a^2c + ac^2 + b^2c + bc^2) + 6abc \tag{24}$$

Using this we find that

$$\Delta_1 = (x_1 + wx_2 + w^2x_3)^3 = x_1^3 + x_2^3 + x_3^3$$
$$+ 3[x_1^2wx_2 + x_1(wx_2)^2 + x_1^2w^2x_3 + x_1(w^2x_3)^2 + (wx_2)^2w^2x_3$$
$$+ (wx_2)(w^2x_3)^2] + 6x_1x_2x_3$$
$$= x_1^3 + x_2^3 + x_3^3 + 3w[x_1^2x_2 + x_1x_3^2 + x_2^2x_3] + 3w^2[x_1x_2^2 + x_1^2x_3 + x_2x_3^2]$$
$$+ 6x_1x_2x_3$$

and, similarly, we find that

$$\Delta_2 = (x_1 + w^2x_2 + wx_3)^3 = x_1^3 + x_2^3 + x_3^3 + 3w^2[x_1^2x_2 + x_1x_3^2 + x_2^2x_3]$$
$$+ 3w[x_1x_2^2 + x_1^2x_3 + x_2x_3^2] + 6x_1x_2x_3$$

Adding and again using the fact that $w^2 + w + 1 = 0$ we find that

$$\Delta_1 + \Delta_2 = 2(x_1^3 + x_2^3 + x_3^3) - 3(x_1^2x_2 + x_1x_2^2 + x_1^2x_3 + x_1x_3^2 + x_2^2x_3 + x_2x_3^2)$$
$$+ 12x_1x_2x_3$$

Finally, using (19), (23), and (17) we get

$$\Delta_1 + \Delta_2 = (x_1 + wx_2 + w^2x_3)^3 + (x_1 + w^2x_2 + wx_3)^3 \tag{25}$$
$$= 2(-p^3 + 3pq - 3r) - 3(-pq + 3r) - 12r$$
$$= -2p^3 + 9pq - 27r$$

Equation (11) is then

$$\overline{X}^2 - [\Delta_1 + \Delta_2]\overline{X} + \Delta_1\Delta_2 = \overline{X}^2 - [-2p^3 + 9pq - 27r]\overline{X} + (p^2 - 3q)^3 = 0$$

From this we can solve the cubic as we showed earlier.

Note. In applying this method we must choose cube roots of Δ_1 and Δ_2. We must be careful to do that in such a way that $\sqrt[3]{\Delta_1 \Delta_2} = p^2 - 3q$.

To illustrate how this method may be used, we shall solve an equation with known roots. We take

$$x^3 - 6x^2 + 11x - 6 = 0$$

whose roots are easily seen to be 1, 2, and 3. Clearly, $p = -6$, $q = 11$, and $r = -6$. Thus

$$\Delta_1 \Delta_2 = (p^2 - 3q)^3 = [36 - 3(11)]^3 = 27$$

and

$$\begin{aligned}\Delta_1 + \Delta_2 &= -2p^3 + 9pq - 27r \\ &= -2(-6)^3 + 9(-6)(11) - 27(-6) = 0\end{aligned}$$

Hence our quadratic is simply

$$\overline{X}^2 + 27 = 0$$

and, solving this, we find that

$$\Delta_1 = i\, 3^{3/2}, \qquad \Delta_2 = -i\, 3^{3/2}$$

We must take the cube root of each of these quantities. Of the three cube roots of i, we select $-i$. With this choice we find that

$$\sqrt[3]{\Delta_1} = -i\sqrt{3} \quad \text{and} \quad \sqrt[3]{\Delta_2} = i\sqrt{3}$$

Note that $\sqrt[3]{\Delta_1 \Delta_2} = -i^2 3 = 3 = p^2 - 3q$. The roots of our equation, x_1, x_2, x_3, are given by

$$\begin{pmatrix} x_1 \\ x_2 \\ x_3 \end{pmatrix} = \frac{1}{3} \begin{pmatrix} 1 & 1 & 1 \\ 1 & w & w^2 \\ 1 & \overline{w}^2 & \overline{w} \end{pmatrix} \begin{pmatrix} -p \\ \sqrt[3]{\Delta_1} \\ \sqrt[3]{\Delta_2} \end{pmatrix}$$

which leads to (taking $w = \frac{-1}{2} + i\frac{\sqrt{3}}{2}$)

$$x_1 = \frac{1}{3}(-p + \sqrt[3]{\Delta_1} + \sqrt[3]{\Delta_2}) = \frac{1}{3}[-(-6) - i\sqrt{3} + i\sqrt{3}] = 2$$

$$x_2 = \frac{1}{3}(-p + w^2\sqrt[3]{\Delta_1} + w\sqrt[3]{\Delta_2}) = \frac{1}{3}[6 - w^2 i\sqrt{3} + wi\sqrt{3}]$$
$$= \frac{1}{3}[6 + i\sqrt{3}(w - w^2)] = \frac{1}{3}[6 + (i\sqrt{3})(i\sqrt{3})] = \frac{1}{3}[6 - 3] = 1$$

$$x_3 = \frac{1}{3}(-p + w\sqrt[3]{\Delta_1} + w^2\sqrt[3]{\Delta_2}) = \frac{1}{3}[6 + w(-i\sqrt{3}) + w^2(\sqrt{3})]$$
$$= \frac{1}{3}[6 + (w^2 - w)i\sqrt{3}] = \frac{1}{3}[6 - (i\sqrt{3})(i\sqrt{3})] = \frac{1}{3}(9) = 3$$

Exercises 7

1. Use Lagrange's method to solve:
   (a) $x^3 - 1 = 0$
   (b) $x^3 - 5x^2 + 8x - 4 = 0$

2. A permutation of a finite set is a one-to-one map from this set onto itself. There are six permutations of the set $\{1, 2, 3\}$. These may be written as

$$a = \begin{pmatrix} 1 & 2 & 3 \\ 2 & 3 & 1 \end{pmatrix}, \quad b = \begin{pmatrix} 1 & 2 & 3 \\ 3 & 2 & 1 \end{pmatrix}, \quad c = \begin{pmatrix} 1 & 2 & 3 \\ 1 & 3 & 2 \end{pmatrix}$$

$$d = \begin{pmatrix} 1 & 2 & 3 \\ 2 & 1 & 3 \end{pmatrix}, \quad e = \begin{pmatrix} 1 & 2 & 3 \\ 1 & 2 & 3 \end{pmatrix}, \quad f = \begin{pmatrix} 1 & 2 & 3 \\ 3 & 1 & 2 \end{pmatrix}$$

where we understand that $a(1) = 2$, $a(2) = 3$, $a(3) = 1$, etc. Let us call $x_1 + wx_2 + w^2 x_3$, y, and let $a(y) = x_{a(1)} + wx_{a(2)} + w^2 x_{a(3)} = x_2 + wx_3 + w^2 x_1$.

   (a) Show that $a(y) = w^2 y$, and $f(y) = wy$.
   (b) Show that $b(y) = w^2 c(y)$, and $d(y) = wc(y)$.
   (c) Conclude that $y^3 = (x_1 + wx_2 + w^2 x_3)^3$ is unaffected by a, e, and f whereas b, c, and d transform it into $(x_1 + w^2 x_2 + wx_3)^3$.

3. Let $\Delta = (x_1 - x_2)(x_1 - x_3)(x_2 - x_3)$. Referring to problem 2, take $a(\Delta) = [x_{a(1)} - x_{a(2)}][x_{a(1)} - x_{a(3)}][x_{a(2)} - x_{a(3)}]$ and similarly for the other permutations. Which permutations map Δ to Δ (these are called even) and which map Δ to $-\Delta$ (these are called odd)?

8. Some Algebraic Matters

We have seen that, for a fixed natural number n, the multiplicative group $(U_n, \cdot)$ is generated by any primitive nth root of unity. In particular, if ω is a primitive 6th root of unity, then $\omega^6 = 1$, six is the smallest positive integer for which this is true, and $U_6 = \{\omega^0, \omega, \omega^2, \omega^3, \omega^4, \omega^5\}$. It is easy to see that ω^2, which is a 6th root of unity, is also a cube root of unity. The same is true of ω^4. The element ω^3 is a square root of unity, whereas ω^5 is primitive.

There is a problem here that we shall solve in this section. A complex number z is an nth root of unity if $z^n = 1$. But, for such a number z, we may have $z^k = 1$ for $k < n$. The smallest positive integer for which this is true was called the order of z (Section 6, Definition 5), $\theta(z)$, and we have seen that $\theta(z)$ must be a divisor of n. What we shall do is figure out how to compute the order of every nth root. We have already seen how to begin. We choose a primitive nth root ω and note that because U_n is equal to $\{\omega^j \mid 0 \leq j \leq n-1\}$, every nth root of unity is equal to ω^j for some $j = 0, 1, 2, \cdots, n-1$. So all we need do is figure out how to compute the order of ω^j. The problem can be solved by recalling a little number theory.

When a, b are integers we say a is a factor, or a divisor, of b if there is an integer m such that $am = b$. When this is the case we say a divides b (or divides evenly into b), and write $a|b$. A useful observation is this:

Lemma 1. If $a, b, c,$ and d are integers, if $a + b = c$, and if d divides any two of the numbers a, b, c, then it divides the third one.

Definition 1. Let a, b be two positive integers. A positive integer d is said to be the greatest common divisor of a and b if (i) $d|a$ and $d|b$; (ii) any integer c that divides both a and b also divides d.

So, for example, the numbers 18 and 24 have 1, 2, 3, and 6 as common divisors. The greatest of these is, of course, 6.

Lemma 2. Any two positive integers a, b have a unique greatest common divisor d. Furthermore, there are integers x_0 and y_0 such that $d = ax_0 + by_0$.

Proof. Let $M(a, b) = \{ax + by \mid x, y \text{ are in } Z\}$. Clearly, $a = a \cdot 1 + b \cdot 0$ and $b = a \cdot 0 + b \cdot 1$ are in $M(a, b)$, and because $(ax_1 + by_1) - (ax_2 + by_2) = a(x_1 - x_2) + b(y_1 - y_2)$, this set is an additive group (Section 4, Definition 1). Hence, by Corollary 2 to Lemma 1 (Section 4), there is a smallest positive integer $d \in M(a, b)$ and $M(a, b) = \{kd \mid k \in Z\}$. It follows that $d|a$ and $d|b$. However, because $d \in M(a, b)$, there are integers x_0 and y_0 such that $d = ax_0 + by_0$. From this, and Lemma 1, we see that any common divisor of a and b must divide d.

To see that d is unique we simply note that any two greatest common divisors of a and b must divide each other. Because they are positive integers they must then be equal.

Definition 2. Given two positive integers a and b, we denote their greatest common divisor (their g.c.d.) by (a, b). We shall say that a and b are relatively prime when $(a, b) = 1$.

From Lemma 2 we see that when $(a, b) = 1$ there are integers x_0 and y_0 such that $1 = ax_0 + by_0$. The converse is also true. If, for two natural numbers a and b, there are integers x_0 and y_0 such that $1 = ax_0 + by_0$, then, by Lemma 1, a and b must be relatively prime. This has a useful consequence. Suppose $\ell = (a, b)$, so $\ell = ax + by$ for some integers x and y. Then $1 = (\frac{a}{\ell})x + (\frac{b}{\ell})y$, showing that the natural numbers $\frac{a}{\ell}$ and $\frac{b}{\ell}$ are relatively prime.

If a, b, c are natural numbers and $a|bc$ we *cannot* conclude that $a|b$ or that $a|c$. For example, $6|24$ so $6|8 \cdot 3$, but 6 does not divide either 8 or 3. This observation gives meaning to our next result.

Lemma 3. Let a, b, c be natural numbers and suppose that a and b are relatively prime. If $a|bc$, then $a|c$.

Proof. We have $1 = (a, b)$ and so $1 = ax + by$ for some integers x and y. Thus $c = (ac)x + (bc)y$ and hence, by Lemma 1, $a|c$.

Recall that a prime number is an integer greater than 1 whose only factors are itself and 1. If p is a prime and a is a natural number, then either $p|a$ or $(p, a) = 1$. Thus:

Corollary 1. Let a, b be natural numbers, let p be a prime, and suppose that $p|ab$. Then either $p|a$ or $p|b$.

We are now ready to compute the order of ω^k, where ω is a primitive nth root of unity and $0 \leq k < n-1$. Recall that the order of ω^k is the smallest positive integer d such that $(\omega^k)^d = 1$. We have seen that if for some positive integer m, $(\omega^k)^m = 1$, then $d|m$. It is easy enough to repeat the argument here. Because $d \leq m$, $m = qd + r$ for integers q and r with $0 \leq r < d$. But $(\omega^k)^m = 1$ and $(\omega^k)^d = 1$ so $(\omega^k)^r = 1$, showing r must be zero.

Theorem 1. Let ω be a primitive nth root of unity. Then the order of ω^k is $\frac{n}{(k,n)}$; i.e., the integer obtained by dividing n by the g.c.d. of n and k. In particular, ω^k is primitive if, and only if, n and k are relatively prime.

Proof. Suppose that the order of ω^k is d and that $\ell = (k, n)$. We want to show that $d = \frac{n}{\ell}$. Because these are positive integers we need only show that they divide each other.

Now $(\omega^k)^{n/\ell} = (\omega^n)^{k/\ell} = (1)^{k/\ell} = 1$ and so, by the observation we made immediately above, $d|(\frac{n}{\ell})$. Because $(\omega^k)^d = 1$, $n|kd$. It follows that $(\frac{n}{\ell})|(\frac{kd}{\ell})$, which shows $(\frac{n}{\ell})|(\frac{k}{\ell}) \cdot d$. But $\frac{n}{\ell}$ and $\frac{k}{\ell}$ are relatively prime. Thus $(\frac{n}{\ell})|d$ and we are done.

The number theory we have discussed here has some further interesting applications that we shall now explore. We first note that any additive subgroup of Z has an unusual property. Given a subgroup (ℓ) of Z we have $(a \cdot b) \in (\ell)$ whenever $b \in (\ell)$, and this is true for all $a \in Z$. It is this property that enables us to introduce a new operation, a kind of multiplication, onto the set Z_ℓ. First we need:

Lemma 4. Let (ℓ) be any additive subgroup of Z and suppose that $a \equiv a'$ mod (ℓ), $b \equiv b'$ mod (ℓ). Then, $a \cdot b \equiv a' \cdot b'$ mod (ℓ).

Proof. Because $a \equiv a'$ mod (ℓ), we have $a = a' + k\ell$ for some integer k. Also, $b = b' + m\ell$ for some integer m. Hence $a \cdot b = a' \cdot b' + [(a'm)\ell + (b'k)\ell + (mk\ell)\ell]$, which proves our result.

Definition 3. Let ℓ be a fixed natural number greater than 1. For any $\bar{a}, \bar{b}$ in Z_ℓ we define $\bar{a} \cdot \bar{b}$ to be $\overline{(ab)}$; note here that $\bar{a}$ denotes the equivalence class, modulo ℓ, that contains a.

It is clear, from our lemma, that the binary operation $\cdot$ is well defined. Moreover, because $\overline{(ab)} = \overline{(ba)}$ we have $\bar{a} \cdot \bar{b} = \bar{b} \cdot \bar{a}$, because $\overline{[a(bc)]} = \overline{[(ab)c]}$ we have $\bar{a} \cdot (\bar{b} \cdot \bar{c}) = (\bar{a} \cdot \bar{b}) \cdot \bar{c}$, and because $a = a \cdot 1$ we have $\bar{a} = \bar{a} \cdot \bar{1}$.

Let us compute the multiplication tables in two cases:

$\cdot$	0	1	2	3	4
0	0	0	0	0	0
1	0	1	2	3	4
2	0	2	4	1	3
3	0	3	1	4	2
4	0	4	3	2	1

$(Z_5, \cdot)$

$\cdot$	0	1	2	3	4	5
0	0	0	0	0	0	0
1	0	1	2	3	4	5
2	0	2	4	0	2	4
3	0	3	0	3	0	3
4	0	4	2	0	4	2
5	0	5	4	3	2	1

$(Z_6, \cdot)$

These tables display some interesting features of this new operation, but before we can discuss them we need some additional terminology. To begin with, the element $\bar{1}$ is called the multiplicative identity of Z_ℓ. An element $\bar{b}$ of Z_ℓ is said to be invertible if it has a multiplicative inverse; i.e., an element $\bar{a} \in Z_\ell$ such that $\bar{a} \cdot \bar{b} = \bar{b} \cdot \bar{a} = \bar{1}$.

The table for $(Z_5, \cdot)$ shows that every nonzero element of this set has a multiplicative inverse. This is not true in Z_6. There we see that only $\bar{1}$ and $\bar{5}$ have inverses, whereas $\bar{2}, \bar{3}, \bar{4}$ do not. Even stranger things happen in $(Z_6, \cdot)$.

The table shows that $\bar{2} \cdot \bar{3} = \bar{0}$ and that $\bar{3} \cdot \bar{4} = \bar{0}$, while none of the factors is zero. We have developed the number theory we need to explain these "weird" results. Again we must start with some terminology.

Definition 4. Two elements $\bar{a}, \bar{b}$ of Z_ℓ are said to be proper divisors of zero if $\bar{a} \neq \bar{0} \neq \bar{b}$, and $\bar{a} \cdot \bar{b} = \bar{0}$.

We observe that a proper divisor of zero could not have a multiplicative inverse (see Exercises 8, problem 2).

Theorem 2. The set $(Z_\ell, \cdot)$ contains proper divisors of zero if, and only if, ℓ is a composite integer (i.e., an integer greater than one that is *not* a prime).

Proof. Suppose first that ℓ is composite, so $\ell = k \cdot n$ for $1 < k < \ell$ and $1 < n < \ell$. Clearly then, $\bar{k} \neq \bar{0} \neq \bar{n}$ whereas $\overline{(kn)} = \bar{0}$, showing that Z_ℓ contains proper divisions of zero.

Now suppose that there are proper divisions of zero in Z_ℓ. Let us say $\bar{k}, \bar{n}$ are nonzero, yet $\bar{k} \cdot \bar{n}$ is zero. Then $kn = m\ell$ for some integer m. If ℓ were a prime it would follow that $\ell | k$, giving us $\bar{k} = \bar{0}$, or $\ell | n$ and implying that $\bar{n} = \bar{0}$. Thus ℓ must be a composite integer.

Corollary 1. When ℓ is a prime the set Z_ℓ does not contain proper divisors of zero.

Our next result tells us which elements of Z_ℓ have, and which do not have, multiplication inverses. Its proof also shows the connection between the latter elements and the proper divisors of zero.

Theorem 3. An element $\bar{k}$ of Z_ℓ has a multiplicative inverse if, and only if, the integers k and ℓ are relatively prime.

Proof. Suppose that $(k, \ell) = 1$. Then there are integers x and y such that $1 = kx + \ell y$. Thus $kx \equiv 1 \mod (\ell)$ showing that $\bar{x}$ is the multiplicative inverse of $\bar{k}$.

Next we suppose that k and ℓ are not relatively prime; i.e., $(k, \ell) = n > 1$. Then $k = k_1 n$ and $\ell = \ell_1 n$ where $1 < k_1 < k$ and $1 < \ell_1 < \ell$. It follows that $k \cdot \ell_1 = (k_1 n) \cdot \ell_1 = k_1 (n \ell_1) = k_1 \ell \in (\ell)$. Thus $\bar{k}$ and $\bar{\ell}_1$ are proper divisors of zero.

Remark 1. The proof of Theorem 3 shows that when $\bar{k} \in Z_\ell$, $\bar{k} \neq \bar{0}$ and $\bar{k}$ is not invertible, then $\bar{k}$ is a proper divisor of zero. More precisely, there is an $\bar{\ell}_1 \in Z_\ell$ such that $\bar{k}$ and $\bar{\ell}_1$ are proper divisors of zero.

Corollary 1. When ℓ is a prime, every nonzero member of Z_ℓ has a multiplicative inverse. Moreover, for $\bar{a}, \bar{b}$ in Z_ℓ we have $\bar{a} \cdot \bar{b} = \bar{0}$ if, and only if, one of these factors is zero.

Remark 2. Given a natural number m let $\mathcal{E}(m) = \{k \in \mathbb{N} \mid k \leq m$ and $(k,m) = 1\}$. Euler defined a function $\varphi = \mathbb{N} \to \mathbb{N}$ by setting, for each $m \in \mathbb{N}$, $\varphi(m)$ equal to the number of elements in the set $\mathcal{E}(m)$. This is called the Euler phi function or, sometimes, the Euler totient function. We shall study it further after we recall some facts about prime numbers.

Exercises 8

1. Let ω be a primitive nth root of unity. Compute the order of each of the numbers ω^j, $j = 0, 1, \ldots, n-1$ when:
   (a) $n = 6$;
   (b) $n = 12$;
   (c) n is a prime number.

∗2. Suppose that $\bar{a}$ and $\bar{b}$ are proper divisors of zero in Z_ℓ. Show that neither of these elements can have a multiplicative inverse in Z_ℓ.

∗3. Suppose that the integers k, a, b and n satisfy $ka \equiv kb \mod n$. If $(k,n) = 1$, show that $a \equiv b \mod n$. Hint: In Z_n we have $\bar{k}\bar{a} = \bar{k}\bar{b}$.

4. Compute $\varphi(1)$, $\varphi(2)$, $\varphi(3)$, $\varphi(4)$, and $\varphi(p)$ where p is a prime.
   (a) If p is not a prime show that $\bar{1} \cdot \bar{2} \cdot \ldots \cdot \overline{(p-1)} = \bar{0}$ in Z_p.
   (b) If p is an odd prime note that $\overline{(p-1)}$ is its own inverse in Z_p. Conclude that $\bar{1} \cdot \bar{2} \cdot \ldots \cdot \overline{(p-2)} = \bar{1}$ in Z_p.
   (c) Use (a) and (b) to show that p is a prime if, and only if, $\bar{1} \cdot \bar{2} \cdot \ldots \cdot \overline{(p-1)} \equiv -1 \mod p$. This is called Wilson's theorem.

9. Prime Numbers

Primes have been investigated since ancient times. A great deal is known about them, and yet many mysteries remain. To begin our discussion let us restrict our attention to the set $\mathbb{N} = \{1, 2, 3, \ldots\}$ of natural numbers.

Given n and d in $\mathbb{N}$ we shall say that d is a divisor, or a factor, of n, and we shall write $d|n$, if there is an $m \in \mathbb{N}$ such that $dm = n$. Observe that any factor of d is also a factor of n.

Definition 1. A natural number $n > 1$ is said to be a prime if the only divisors of n are n and 1.

Clearly 2, 3, 5, 7, 11, 13, 17, 19, 23, 29, 31, ... are primes. A number $n > 1$ that is not a prime is called a composite number.

For any $n \in \mathbb{N}$ let $D(n) = \{d \in \mathbb{N} |\, d|n\}$, so $D(n)$ is the set of all factors of n. When $n > 1$ the set $D(n) \backslash \{1\} \neq \varnothing$, and so it has a smallest member p. Observe that p must be a prime; for if $1 < q < p$ and $q|p$, then $q|n$ is smaller than the smallest divisor of n. Thus every $n > 1$ has a prime factor.

Theorem 1. There are infinitely many prime numbers.

Proof. Suppose that the set of primes is finite and let this set be denoted by $p_1, p_2, \ldots, p_k$, $k \in \mathbb{N}$. Set

$$Q = p_1 \cdot p_2 \cdot \cdots \cdot p_k + 1$$

and note that no p_j can divide Q; for, if it did, then by Lemma 1 of Section 8, it would have to divide 1. But then any prime factor of Q, and there is at least one, is not on our list of all primes, an obvious contradiction.

If n is a composite number, then $n = p_1 q_1$, where p_1 is the smallest prime divisor of n. If q_1 is a prime, then n is the product of two primes. If q_1 is composite, then $q_1 = p_2 q_2$, where p_2 is the smallest prime divisor of q_1. Thus $n = p_1 p_2 q_2$. Continuing in this way we should be able to write n as a product of prime numbers. This is, in fact, the case as we shall now prove.

Theorem 2. Any integer greater than 1 is either a prime or it can be written as a product of primes.

Proof. We argue by contradiction. Let M be the set of natural numbers greater than 1 that are not prime and cannot be written as a product of primes. Suppose $M \neq \varnothing$. Then there is a smallest element m of M. Clearly m is not a prime and so $m = ab$, where $1 < a < m$, $1 < b < m$. But because $a < m$ and $b < m$, neither of these numbers is in M. Thus they are primes or can be written as products of primes. But then the integer m can be written as a product of primes, and we have reached a contradiction.

It seems natural, given $n > 1$, to ask if there is more than one way to factor n into primes. It turns out that apart from permuting the factors, this can be done in just one way.

Theorem 3. Given $n > 1$ we may find primes $p_1 < p_2 < \cdots < p_k$ and natural numbers $a_1, a_2, \cdots, a_k$ such that

$$n = p_1^{a_1} p_2^{a_2} \cdots p_k^{a_k}$$

This can be done in one, and only one, way.

Proof. Assume that the theorem is false. Then there is a smallest integer $m > 1$ that has two distinct factorizations: $m = p_1^{a_1} \cdots p_k^{a_k}$, where $p_1 < p_2 < \cdots < p_k$ are primes and $a_1, \ldots, a_k$ are natural numbers, and $m = q_1^{b_1} \cdots q_\ell^{b_\ell}$ where $q_1 < \cdots < q_\ell$ are primes and $b_1, \ldots, b_\ell$ are natural numbers.

It follows that $p_1 | q_1^{b_1} \cdots q_\ell^{b_\ell}$ and so $p_1 | q_i^{b_i}$ for some i, $1 \leq i \leq \ell$ (Section 8, Corollary 1 to Lemma 3). From this we see that $p_1 | q_i$ and $p_1 = q_i$. A similar argument shows that $q_1 = p_j$ for some j, $1 \leq j \leq k$. Thus

$$p_1 \leq p_j = q_1 \leq q_i = p_1$$

giving us $p_1 = q_1$.

Now $\frac{m}{p_1} < m$ and so this number (i.e., $\frac{m}{p_1}$) can be written, as described above, in just one way. But

$$\frac{m}{p_1} = p_1^{a_1-1} p_2^{a_2} \cdots p_k^{a_k} = q_1^{b_1-1} a_2^{b_2} \cdots q_\ell^{b_\ell}$$

It follows that $k = \ell$, $p_2 = q_2$, ..., $p_k = q_k$, and $a_1 - 1 = b_1 - 1$, $a_2 = b_2, \ldots, a_k = b_k$. Thus the factorization of m is unique, and we have reached a contradiction.

For any $n \in \mathbb{N}$ let us set $\sigma(n)$ equal to the sum of all the divisors of n. We observe that

$$\sigma(7) = 8 < 2 \cdot 7, \quad \sigma(12) = 28 > 2 \cdot 12, \quad \sigma(6) = 12 = 2(6)$$

The number n is called deficient when $\sigma(n) < 2n$, it is called abundant when $\sigma(n) > 2n$, and it is said to be perfect when $\sigma(n) = 2n$. This terminology goes back to the numerologist of antiquity.

Perfect numbers were discussed by Euclid, and he gave a means of generating them. He considered numbers of the form

$$n = 2^{p-1}(2^p - 1)$$

When $2^p - 1$ is a prime, the factors of n are

$$1, 2, 2^2, \ldots, 2^{p-1}, \quad 1 \cdot (2^p - 1), \quad 2 \cdot (2^p - 1), \ldots, \quad 2^{p-1} \cdot (2^p - 1)$$

and so

$$\begin{aligned}\sigma(n) &= 1 + 2 + \cdots + 2^{p-1} + (2^p - 1)(1 + 2 + \cdots + 2^{p-1}) \\ &= (2^p - 1) + (2^p - 1)(2^p - 1) \\ &= (2^p - 1)[1 + (2^p - 1)] = 2^p(2^p - 1) = 2n\end{aligned}$$

To sum up these observations: Whenever $2^p - 1$ is a prime, the number $2^{p-1}(2^p - 1)$ is perfect.

We turn now to the question of when $2^p - 1$ is prime. It is easy to see that when n is composite so also is $2^n - 1$ (see Exercises 9, problem 5). Thus we need only consider the numbers

$$M_p = 2^p - 1$$

where $p = 2, 3, 5, 7, 11, \ldots$. These numbers were discussed by the French cleric Mersenne and have come to be called Mersenne numbers. Note that $M_2 = 2^2 - 1 = 3$, $M_3 = 2^3 - 1 = 7$, $M_5 = 2^5 - 1 = 31$, $M_7 = 2^7 - 1 = 127$ are all primes. Unfortunately, $M_{11} = 2^{11} - 1 = 2047 = (23)(89)$. Still the first four Mersenne numbers give us four perfect numbers: $2^{2-1}(3) = 6$, $2^{3-1}(7) = 28$, $2^{5-1}(31) = 496$, and $2^{7-1}(127) = 8128$.

In 1876 Lucas showed that

$$M_{127} = 170,141,183,460,469,231,731,687,303,715,884,105,727$$

is a prime. It is now known that M_{11213}, M_{19737}, M_{21701}, and M_{23209} are primes.

Given a prime p the Mersenne number M_p may, or may not, be a prime. It is useful then to have some test, especially one that can be easily programmed on a computer, to determine when M_p is a prime. Such a test was discovered by Lucas and put in final form by Lehmer. He showed that M_p is a prime if, and only if, $M_p | S_{p-1}$ where the numbers S_n are defined as follows: $S_1 = 4$, $S_{n+1} = S_n^2 - 2$, $n = 1, 2, 3, \ldots$.

We shall give a proof of the "interesting" half of this result; i.e., that when $M_p | S_{p-1}$, M_p is a prime. A few preliminary results are needed here.

Lemma 1. Let $\alpha = 2 + \sqrt{3}$, $\overline{\alpha} = 2 - \sqrt{3}$. Then $\alpha\overline{\alpha} = 1$ and $S_m = \alpha^{2^{m-1}} + \overline{\alpha}^{2^{m-1}}$.

Proof. When $m = 1$ we have $S_1 = \alpha + \overline{\alpha}$, which is clearly true. Assume that there is a smallest integer $k + 1$ for which our formula fails. Then, for $m = 1, 2, \ldots, k$ the formula is true. Thus

$$S_{k+1} = S_k^2 - 2 = \left(\alpha^{2^{k-1}} + \overline{\alpha}^{2^{k-1}}\right)^2 - 2$$

$$= \left(\alpha^{2^{k-1}}\right)^2 + 2(\alpha\overline{\alpha})^{2^{k-1}} + \left(\overline{\alpha}^{2^{k-1}}\right)^2 - 2$$

$$= \alpha^{2^k} + \overline{\alpha}^{2^k}$$

because $\alpha\overline{\alpha} = 1$. This is a contradiction and the lemma is proved.

Suppose now that M_p divides S_{p-1}. Then M_p must divide $\alpha^{2^{p-2}} + \overline{\alpha}^{2^{p-2}}$ and so this sum is equal to RM_p for some integer R. Multiplying by $\alpha^{2^{p-2}}$ gives us

$$\alpha^{2^{p-1}} = RM_p \alpha^{2^{p-2}} - 1 \tag{1}$$

and squaring this gives us

$$\alpha^{2^p} = (RM_p \alpha^{2^{p-2}} - 1)^2 = KM_p + 1 \qquad (2)$$

for some integer K. Assume that M_p is a composite number. Then M_p has a prime factor q with $q^2 \leq M_p$ (see Exercises 9, problem 2). Recalling the definition of M_p we see that $q \neq 2$.

Define a set $\overline{X} = \{a + b\sqrt{3} \mid a, b \in Z_q\}$. Clearly, $\overline{X}$ contains q^2 elements, among them 0 and 1. For x, y in $\overline{X}$ we define $x \cdot y$ as follows: $x = a + b\sqrt{3}$, $y = c + d\sqrt{3}$, $x \cdot y = (ac + 3bd) + (ad + bc)\sqrt{3}$, where the terms $(ac + 3bd)$ and $(ad + bc)$ are found in Z_q. Clearly, $x \cdot y \in \overline{X}$. For any $x \in \overline{X}$ we define $x^1 = x$, $x^2 = x \cdot x$, etc., and we shall say that x is invertible if there is a $y \in \overline{X}$ such that $x \cdot y = 1$. Let Y be the set of invertible elements of $\overline{X}$ and note that Y contains at most $q^2 - 1$ elements.

Lemma 2. For each $y \in Y$, $y^r \equiv 1$ for some natural number $r \leq q^2 - 1$.

Proof. If $y \in Y$, then each of the elements $y, y^2, y^3, \ldots, y^{q^2-1}$ is in Y. There are two cases: (i) If the numbers listed are all distinct, then one of them must equal 1 (because Y contains at most $q^2 - 1$ members) and we are done. (ii) If $y^\ell \equiv y^m$ for $1 \leq \ell < m$, then $1 \equiv y^{m-\ell}$, and again we are done.

For each $y \in Y$ let $\mathcal{O}(y)$ be the smallest positive integer such that $y^{\mathcal{O}(y)} \equiv 1$. Then if r is any integer such that $y^r \equiv 1$ we must have $\mathcal{O}(y) \mid r$ (see Section 6, the discussion just below Definition 5).

Now $\alpha \in \overline{X}$ and because $q \mid M_p$, $\alpha^{2^{p-1}} \equiv -1 \mod q$ by (1) and $\alpha^{2^p} \equiv 1 \mod q$ by (2). Thus $\alpha \in Y$ and $\mathcal{O}(\alpha) \mid 2^p$, and $\mathcal{O}(\alpha)$ cannot be less than 2^p because $\alpha^{2^{p-1}} \equiv -1$. (Note that because $q > 2$ we do *not* have $1 \equiv -1$.) Thus

$$\mathcal{O}(\alpha) = 2^p \leq q^2 - 1 \leq M_p - 1 = 2^p - 2$$

and we have reached a contradiction.

Exercises 9

1. If $d \mid n$ and $r \mid d$, show that $r \mid n$.

* 2. If n is a composite number and p is the smallest prime factor of n, show that $p^2 \leq n$.

3. Factor the following numbers into primes:

    (a) 1001
    (b) 756

(c) At some point in his life, Euler became blind. It was after that that he factored 1,000,009 into (293)(3413).

4. For any $n > 1$ consider $\sum \frac{1}{d}$ where the sum is over all divisors d of n. Show that this sum is less than 2 when n is deficient, greater than 2 when n is abundant, and equal to 2 when n is perfect. Hint: Compute $\sum \frac{n}{d}$.

*5. When n is composite, show that $2^n - 1$ is composite.

6. Euler showed that every even perfect number is the form $2^{p-1} M_p$, where M_p is prime. At the time of this writing, it is not known if odd perfect numbers exist.

10. Euler's Phi Function

Given $m \in \mathbb{N}$ the number of invertible elements of Z_m coincides with the number of integers k such that $1 \leq k \leq m$ and $(k, m) = 1$. This is also the number of primitive mth roots of unity. As we have seen (Section 8, Remark 2) the function $\varphi(m)$ gives us this number. Our purpose here is to derive a formula for computing $\varphi(m)$ from m.

Lemma 1. Fix $b \in \mathbb{N}$ and consider the set Z_b. It consists of the equivalence classes $\bar{0}, \bar{1}, \ldots, \overline{b-1}$. Suppose that $r \geq 0$ is an integer and $a \in \mathbb{N}$ satisfies $(a, b) = 1$. Then $\bar{r}, \overline{(a+r)}, \overline{(2a+r)}, \ldots, \overline{((b-1)a+r)}$ are all the elements of Z_b.

Proof. The list given contains b elements of Z_b. It suffices only to show that these elements are distinct. Now if $sa + r \equiv ta + r \mod b$, where $0 \leq s, t \leq b-1$, then $sa \equiv ta \mod b$. However, because $(a, b) = 1$ this tells us that $s \equiv t \mod b$ (Exercises 8, problem 3), and this in turn, because $0 \leq s, t \leq b - 1$ gives us $s = t$.

To illustrate the power of Lemma 1, we use it to prove a result that has found application in securing computer files.

Theorem 1 (Fermat). Let a, p be natural numbers and suppose that p is a prime. If p does not divide a, then $a^{p-1} \equiv 1 \mod p$.

Proof. Because p does not divide a it cannot divide any of the numbers $a, 2a, 3a, \ldots, (p-1)a$. By Lemma 1, with $r = 0$, the set $\bar{a}, \overline{2a}, \overline{3a}, \ldots, \overline{p-1a}$ and $\bar{0}$, coincides with Z_p. It follows that $\bar{a}, \overline{2a}, \ldots, \overline{(p-1)a}$ coincides with the set $\bar{1}, \ldots, \overline{p-1}$. Thus, in Z_p, we have

$$\overline{a^{p-1} \cdot 1 \cdot 2 \cdots (p-1)} = \overline{1 \cdot 2 \cdots (p-1)}$$

or, equivalently, by

$$a^{p-1} \cdot 1 \cdot 2 \cdots (p-1) \equiv 1 \cdot 2 \cdots (p-1) \mod p$$

But $(1 \cdot 2 \cdots (p-1),\ p) = 1$ and so by Exercises 8, problem 3, we must have

$$a^{p-1} \equiv 1 \mod p.$$

Corollary 1. If p is a prime and a is a natural number, then $a^p \equiv a \mod p$.

Let us return now to our discussion of the phi function.

Theorem 2. If a, b are natural numbers and $(a,b) = 1$, then $\varphi(ab) = \varphi(a)\varphi(b)$.

Proof. Consider an integer k, $0 \le k < ab$. We may write $k = aq + r$, where $0 \le r \le a-1$. But note that we also have $0 \le q \le b-1$. In fact, each k, $0 \le k < ab$ can be written in one, and only one, way as $k = aq + r$ with q and r as given above.

Now given k, $0 \le k < ab$ we ask if $(k, a) = 1$. This may be written $(aq + r,\ a) = 1$, and this is true if, and only if, $(r, a) = 1$. There are $\varphi(a)$ numbers r that are smaller than a and prime to a. Choose and fix one of them, say r_1.

By Lemma 1 the list $r_1, a+r_1, 2a+r_1, \ldots, (b-1)a+r_1$, contains b numbers distinct mod b. Hence this list contains exactly $\varphi(b)$ numbers that are prime to b.

Thus each of the $\varphi(a)$ numbers r gives us $\varphi(b)$ numbers that are prime to both a and b. But the number of integers that are prime to both a and b is $\varphi(ab)$ and, as we have just seen, there are $\varphi(a)\varphi(b)$ such numbers.

Corollary 1. If p is a prime and n is a natural number, then $\varphi(p^n) = p^n(1 - \frac{1}{p})$.

Proof. The natural numbers that are $\le p^n$ and *not* prime to p^n are $p, 2p, 3p, \ldots, p^{n-1}p$. There are then, p^{n-1} such numbers. Hence the number of natural numbers that are $\le p^n$ and prime to p^n must be $p^n - p^{n-1} = p^n(1 - \frac{1}{p})$.

We have seen that any natural number $m > 1$ is either a prime or it can be written as a product of primes in "essentially" only one way. When m is a prime $\varphi(m) = m - 1$. When $m = p_1^{e_1} \cdots p_k^{e_k}$ we have:

Corollary 2. $\varphi(m) = \varphi(p_1^{e_1} \cdots p_k^{e_k}) = m(1 - \frac{1}{p_1}) \cdots (1 - \frac{1}{p_k})$.

Proof. It is clear that $(p_i^{e_i}, p_j^{e_j}) = 1$ when $i \ne j$. Thus

$$\varphi(p_1^{e_1} \cdots p_k^{e_k}) = \varphi(p_1^{e_1}) \cdots \varphi(p_k^{e_k})$$

By Corollary 1 we know that $\varphi(p_j^{e_j}) = p_j^{e_j}(1 - \frac{1}{p_j})$. Hence we see that

$$\varphi(m) = p_1^{e_1}\left(1 - \frac{1}{p_1}\right) \cdot p_2^{e_2}\left(1 - \frac{1}{p_2}\right) \cdots p_k^{e_k}\left(1 - \frac{1}{p_k}\right)$$

$$= p_1^{e_1} \cdot p_2^{e_2} \cdots p_k^{e_k}\left(1 - \frac{1}{p_1}\right) \cdots \left(1 - \frac{1}{p_k}\right)$$

$$= m\left(1 - \frac{1}{p_1}\right) \cdots \left(1 - \frac{1}{p_k}\right)$$

When p, a are natural numbers and p is a prime, the assertion that p does not divide a means $(p, a) = 1$. Fermat showed that we then have $a^{p-1} \equiv 1 \mod p$. Recall that $\varphi(p) = p - 1$ so we have $a^{\varphi(p)} \equiv 1 \mod p$. Euler showed that for any natural number m if $(m, a) = 1$, then $a^{\varphi(m)} \equiv 1 \mod m$. This generalization of Fermat's result also plays a role in securing computerized information. We shall discuss this application further after we prove Euler's result.

Theorem 3. Let a, m be natural numbers and suppose that $(a, m) = 1$. Then $a^{\varphi(m)} \equiv 1 \mod m$.

Proof. Let the $n = \varphi(m)$ natural numbers that are $\leq m$ and prime to m be denoted by $a_1, a_2, \ldots, a_n$. Note that $(aa_j, m) = 1$ for $j = 1, 2, \ldots, n$. Furthermore, if $aa_i \equiv aa_j \mod m$, then $a_i \equiv a_j \mod m$ by Exercises 8, problem 3. Because $1 \leq a_i, a_j \leq m$, this means $a_i = a_j$. Thus the n elements of Z_m, $\overline{aa_1}, \overline{aa_2}, \ldots, \overline{aa_n}$ are distinct and each of them is invertible in Z_m. Hence they must coincide with $\overline{a_1}, \overline{a_2}, \ldots, \overline{a_n}$. But then, in Z_m,

$$(\overline{aa_1}) \cdot (\overline{aa_2}) \cdots (\overline{aa_n}) = \overline{a_1} \cdot \overline{a_2} \cdots \overline{a_n}$$

or, equivalently,

$$a^n \cdot a_1 \cdot a_2 \cdots a_n = a_1 a_2 \cdots a_n \mod m.$$

Finally, because $(a_1 \cdots a_n, m) = 1$ we may again use problem 3 of Exercises 8 to conclude that $a^n \equiv 1 \mod m$.

Let us finish up with a brief discussion of how the theorems of Fermat and Euler are used in connection with the problem of coding and decoding confidential messages. We shall assume that our message must first be converted into an integer M. We must first encode M to obtain the encoded message E. This is then sent to our correspondent who must, somehow, derive M from E.

As a first example, suppose we choose a prime r and a natural number s such that $(s, r - 1) = 1$. Given M we encode it by setting $E = M^s$.

The problem now is to obtain M from E. So we want to find t so that $E^t = M^{st} \equiv M \mod r$. Suppose we can find t so that $st = k(r-1)+1$ for some integer k. Then

$$E^t = (M^s)^t = M^{st} = M^{k(r-1)+1} = (M^{r-1})^k M$$

Because r is a prime, Fermat's theorem tells us that $M^{r-1} \equiv 1 \mod r$ and so $E^t \equiv M \mod r$ as required.

To find t so that $st = k(r-1)+1$, we use Euler's theorem. The equation says that $st \equiv 1 \mod (r-1)$. Now $s^{\varphi(r-1)} \equiv 1 \mod (r-1)$ provided $(s, r-1) = 1$, and we chose s so that this is true. Hence all we need to do is take $t = s^{\varphi(r-1)-1}$. With this choice $st = s^{\varphi(r-1)} \equiv 1 \mod (r-1)$ as required.

As an example, let us take $r = 17$ and $s = 3$. Then $(3, 17-1) = (3, 16) = 1$ as required. Given M we find $E = M^3$ and send this to our correspondent. To decode he must have t such that $st \equiv 1 \mod (r-1)$. We have seen that $t = s^{\varphi(r-1)-1}$. Now $r = 17$ so $r-1 = 16$ and so $\varphi(r-1) = \varphi(16) = 8$. Thus $t = s^7 = 3^7 = 2187$. We may reduce this by noting that $2187 \equiv 11 \mod 16$. Thus to find M we simply compute $E^{11} \mod 17$.

A variant on this technique concerns "public access" codes. Here the r is published openly. It is a many digit number that those using the code know is the product of two very large primes p and q. One chooses s so that $(s, \varphi(r)) = 1$ and encodes the message, M, as before. So $E = M^s$ is sent.

We now seek t so that $E^t = M^{st} \equiv M \mod r$. Again we want $st \equiv 1 \mod \varphi(r)$ or $st = k[\varphi(r)] + 1$ for some k. So we take $t = s^{\varphi(\varphi(r))-1}$. Then $st = s^{\varphi(\varphi(r))} \equiv 1 \mod \varphi(r)$ by Euler's theorem.

Finally, $E^t = M^{st} = M^{k\varphi(r)+1} = [M^{\varphi(r)}]^k M \equiv M \mod r$ because $M^{\varphi(r)} \equiv 1 \mod r$.

Note that because the receiver knows that $r = pq$, s/he also knows $\varphi(r) = (p-1)(q-1)$.

Exercises 10

1. Prove Corollary 1 to Fermat's theorem.

2. Compute $\varphi(m)$ for the following numbers:
   (a) $m = (64)(81)(625)$;
   (b) $m = 2^{10}$;
   (c) $m = 7^2 \cdot 19 \cdot 37^2$.

3. If p is a prime, show that $\varphi(1) + \varphi(p) + \varphi(p^2) + \cdots + \varphi(p^n) = p^n$.

4. There is a conjecture, unproven at the time of this writing, that n is a prime if, and only if, $\varphi(n) \mid (n-1)$.

Notes

1. The axiom of choice, and its many equivalent forms, is discussed in "General Topology" by John L. Kelley (D. Van Nostrand Company, Inc., 1955). See pp. 32–35.

    A proof of statement (∗) used in the discussion of Weiner's theorem (Section 2) can be found in "Introduction to Topology and Modern Analysis" by George F. Simmons (McGraw-Hill Book Company, New York, 1963). See p. 321. See also pp. 308–310 for a further discussion of related ideas.

    Another proof of Wiener's theorem is given in "An Introduction to Harmonic Analysis" by Yitzhak Katznelson (John Wiley & Sons, Inc., New York, 1968). This is done on p. 202 where (∗) is also proved.

2. The Fourier transform on $\mathbb{R}$, using Lebesque integration, is treated in Katznelson's book (Note 1 above) and in the book "Fourier Transforms" by R. R. Goldberg (Cambridge University Press, New York, 1961). Goldberg, in an appendix, discusses the Fourier transform on a locally compact abelian group, and Katznelson has a chapter on this topic. A more extensive discussion can be found in "Abstract Harmonic Analysis" by E. Hewitt and K. A. Ross (Springer-Verlag, New York, 1963).

3. Lagrange's work on the theory of equations is discussed in detail by Harold M. Edwards in his book "Galois Theory" (Springer, New York, 1984). See p. 2 and pp. 18–22. Edwards also discusses some related work of Vandermonde and, on p. 19, gives a brief (and to me at least, surprising) biography of Lagrange: "Unlike Vandermonde, who was French but did not have a French name, Lagrange had a French name but was not French. He was Italian (he was born with the name Lagrangia and his native city was Turin) and at the time of the publication of his Réflexions he was a member of Frederick the Great's Academy in Berlin. When he left Berlin in 1787 he went to Paris, where he spent the rest of his life and where, of course, he was a leading member of the scientific community; this has tended to reinforce the impression that he was French. He was certainly the greatest mathematician of the generation between Euler and Gauss, and, indeed, has a secure place among the great mathematicians of all time."

4. Our definition of the set $\overline{X}$ and its algebra leaves something to be desired (see Chapter Five, Exercises 3, problem 2). The interested reader can also consult the article by J. W. Bruce, "A Really Trivial Proof of the Lucas-Lehmer Test" in *The American Mathematical Monthly*, vol. 100, number 4, April 1993, pp. 370–371.

    Mersenne numbers are useful in digital signal processing, and certain finite Fourier transforms are sometimes called Mersenne number transforms (see 2 in the Notes to Chapter 5).

5. The Euler Phi function is discussed in "Introduction to the Theory of Numbers" by Leonard Eugene Dickson (Dover Publication, Inc., New York, 1957), see pp. 7 and 8, and in "Number Theory in Science and Communication" by M. R. Schroeder (Springer-Verlag, New York, 1986) on pp. 113–116. Schroeder also discusses using the phi function for "Public-Key Encryption" on pp. 118–126.

An application of Euler's function to signal processing is given in a paper by N. K. Bose and Kim in the JEEE Trans. Signal Processing, vol. 39, no. 10, Oct. 1991, pp. 2167–2173.

Chapter Five
Abstract Algebra

Many of the topics discussed earlier in the text can be treated more systematically by bringing in some ideas from modern algebra. To do this, we must raise the level of abstraction a little bit. The gain in insight, however, is well worth the price we must pay to achieve it. In this last chapter, we very briefly discuss groups, rings, algebras, and fields and indicate how they arise in harmonic analysis.

1. Groups

Groups are found in many areas of mathematics—we have seen lots of them already—and also in physics and certain parts of chemistry. One might expect that a concept with such a wide range of applicability would be defined abstractly. This is the case, and it takes some getting used to. To begin with, there is only one binary operation (see below) in a general group. This can be something very familiar like addition or multiplication or it can be something more exotic like convolution or function composition. The examples below make this clear.

When we say that a nonempty set G is closed under a single-valued binary operation, we mean that we have a function $\varphi : G \times G \to G$. It is convenient, and customary, to denote $\varphi(g, h)$ by $g \circ h$ (read "g circle h") for each pair $(g, h) \in G \times G$ and to say that G is closed under the binary operation $\circ$; we shall always assume that $\circ$ is single valued.

Definition 1. Let G be a nonempty set that is closed under a binary operation $\circ$. We shall say that the pair $(G, \circ)$ is a group if (a) for all g, h, k in G we have $g \circ (h \circ k) = (g \circ h) \circ k$; (b) there is an element e in G such that

$g \circ e = e \circ g = g$ for every $g \in G$; (c) for each $g \in G$ there is an element $h \in G$ such that $g \circ h = h \circ g = e$.

We call (a) the associative law and say that the operation $\circ$ is associative. The element e, whose existence is postulated in (b), is called the identity of G. The element h discussed in (c) is called the inverse of the given element $g \in G$ and is usually denoted by g^{-1}; we shall see that, given g, the inverse element is uniquely determined.

The pair $(\mathbb{C},+)$ is clearly a group because we know that addition of complex numbers is an associative operation, that $\mathbb{C}$ has an identity (the number 0), and that each $z \in \mathbb{C}$ has an inverse (the number $(-z)$).

The pair $(\mathbb{C}, \cdot)$ is not a group. It is true that multiplication of complex numbers is associative, and the number 1 is the identity for this operation. However, the number zero, which certainly belongs to $\mathbb{C}$, does not have a multiplicative inverse. It is easily seen, however, that $(\mathbb{C}\setminus\{0\}, \cdot)$ is a group.

Let us show how (a), (b), and (c) of Definition 1 imply that the identity and inverses are unique.

Lemma 1. In any group $(G, \circ)$ there is only one identity element e. Furthermore, given $g \in G$ there is only one element $h \in G$ such that $g \circ h = h \circ g = e$.

Proof. We are assuming that $g \circ e = e \circ g = g$ for all $g \in G$. Suppose that e' is an other element of G for which $g \circ e' = e' \circ g = g$ for all $g \in G$. Then, in particular, $e = e \circ e' = e' \circ e$ and $e' = e' \circ e = e \circ e'$. Because $e \circ e' = e' \circ e$ we see that $e = e'$.

Now let $g \in G$ and suppose we have two elements h, k in G such that $g \circ h = h \circ g = e$ and $g \circ k = k \circ g = e$. Then $h = h \circ e = h \circ [g \circ k] = [h \circ g] \circ k = e \circ k = k$ and we are done.

Corollary 1. For any $g \in G$, $\left(g^{-1}\right)^{-1} = g$.

Proof. Given $g \in G$ there is an element g^{-1} such that $g \circ g^{-1} = g^{-1} \circ g = e$. But this says g is an inverse of g^{-1}. Now the inverse of g^{-1} is, by definition, $\left(g^{-1}\right)^{-1}$. Thus, by Lemma 1, $g = \left(g^{-1}\right)^{-1}$.

Definition 2. Let $(G, \circ)$ be a group and let H be a nonempty subset of G. If $(H, \circ)$ is a group we shall say that H is a subgroup of G.

If H is a subgroup of $(G, \circ)$, then because $H \neq \emptyset$ there is an element $h \in H$. But because $(H, \circ)$ is a group, h^{-1} must belong to H. Furthermore, because H is closed under the operation $\circ$, $h \circ h^{-1} = e$ is in H.

Lemma 2. Let $(G, \circ)$ be a group and let H be a nonempty subset of G. Then H is a subgroup of G if, and only if, $g \circ h^{-1} \in H$ whenever g and h are in H.

Proof. Our discussion just above this lemma shows that our condition is necessary. Suppose now that H satisfies this condition. Then because $H \neq \emptyset$ we can choose $h \in H$ and note that $e = h \circ h^{-1} \in H$. Thus $(H, \circ)$ satisfies (a) and (b) of Definition 1; we have just proved (b), and (a) follows from the fact that $(G, \circ)$ is a group. To prove (c) choose any $h \in H$ and note that because $e \in H$, $h^{-1} = e \circ h^{-1}$ is in H. Finally, we must show that H is closed under $\circ$; i.e., we must show that $g \circ h \in H$ whenever $g, h \in H$. So let us suppose $g, h \in H$ are given. Then $h^{-1} \in H$ as we have seen. Hence $g \circ \left(h^{-1}\right)^{-1} \in H$ and, by Corollary 1, this is just $g \circ h$.

We have seen that $(\mathbb{C}, +)$ is a group. By Lemma 2 then a nonempty subset A of $\mathbb{C}$ is a subgroup of $\mathbb{C}$ if $(a - b) \in A$ whenever a, b are in A. Referring to Chapter Four, Section 4, Definition 1, we see that what we have called an additive group is simply a subgroup of $(\mathbb{C}, +)$.

Similarly, because $(\mathbb{C}\backslash\{0\}, \cdot)$ is a group, a nonempty subset M of $(\mathbb{C}\backslash\{0\})$ is a subgroup if $x \cdot y^{-1} \in M$ whenever x and y are in M. Referring to Definition 2 of Chapter Four, Section 4, we see that what we have called a multiplicative group is simply a subgroup of $(\mathbb{C}\backslash\{0\}, \cdot)$.

Definition 3. A group $(G, \circ)$ is said to be an abelian group, or a commutative group, if the binary operation $\circ$ satisfies the commutative law; i.e., if $g \circ h = h \circ g$ for all $g, h \in G$.

Because addition and multiplication of complex numbers are commutative operations, all subgroups of $(\mathbb{C}, +)$ and of $(\mathbb{C}\backslash\{\circ\}, \cdot)$ are abelian groups.

Recall that given an additive group A and a subgroup C of A we said that elements a, b of A were congruent modulo C, and we wrote $a \equiv b \bmod C$, when $(a - b) \in C$ (Chapter Four, Section 5, Definition 1). We showed that this gives us an equivalence relation on A, and we denoted the set of equivalence classes by A/C. In the case of a multiplicative group M with subgroup N, we defined $s \equiv t \bmod N$ to mean $s \cdot t^{-1} \in N$ (Chapter Four, Exercises 5, problem 1). In this case too we get an equivalence relation on M, and we denoted the set of equivalence classes by M/N. Not surprisingly, we can do this for any group G and any of its subgroups H. We sketch the process here and leave the details, which are almost identical to those in the special cases mentioned above, to the reader.

Given a group $(G, \circ)$ and a subgroup H of G we define for $g, h \in G$, $g \equiv h \bmod H$ to mean that $g \circ h^{-1} \in H$. We say g is congruent to h modulo H.

Lemma 3. Congruence modulo H is an equivalence relation on G.

The set of equivalence classes of G defined by the relation congruence modulo H is denoted by G/H. We have a natural mapping $\varphi_H : G \to G/H$ defined by setting $\varphi_H(g)$ equal to the equivalence class of G (i.e., the element of G/H) that contains g, for each $g \in G$. Calling this equivalence class $[g]$ we have $\varphi_H(g) = [g]$ for each $g \in G$.

Definition 4. Let $(G, \circ)$ be an *abelian* group and let H be a subgroup of g. For each $[g], [h]$ in G/H define $[g] * [h]$ to be $[g \circ h]$.

Lemma 4. Referring to Definition 4 we claim that $(G/H, *)$ is an abelian group.

Proof. We must first show that $*$ is well defined. So suppose we are given $[g], [h]$ in G/H and we choose two elements g', g in $[g]$ and two elements h', h in $[h]$. We must show that $g' \circ h' \equiv g \circ h \bmod H$, for then we have $[g' \circ h'] = [g \circ h]$. Now $g' \circ g^{-1}$ and $h' \circ h^{-1}$ are in H and so their "product," $(g' \circ g^{-1}) \circ (h' \circ h^{-1})$, is in H. But $(g' \circ g^{-1}) \circ (h' \circ h^{-1}) = g' \circ [g^{-1} \circ (h' \circ h^{-1})] = g' \circ [(g^{-1} \circ h') \circ h^{-1}]$, and because $(G, \circ)$ is abelian $g^{-1} \circ h' = h' \circ g^{-1}$. Hence $(g' \circ g^{-1}) \circ (h' \circ h^{-1}) = g' \circ [(h' \circ g^{-1}) \circ h^{-1}] = g' \circ [h' \circ (g^{-1} \circ h^{-1})] = (g' \circ h') \circ (g \circ h)^{-1}$ because $(g \circ h)^{-1} = h^{-1} \circ g^{-1} = g^{-1} \circ h^{-1}$ in this case.

The fact that $*$ is associative and commutative follows from the fact that $\circ$ has these properties. More explicitly,

$[g] * ([h] * [k]) = [g] * [h \circ k] = [g \circ (h \circ k)] = [(g \circ h) \circ k] = [g \circ h] * [k] = ([g] * [h]) * [k]$ and $[g] * [h] = [g \circ h] = [h \circ g] = [h] * [g]$ for all g, h, k in G.

The identity element of G/H is just $[e]$ because $[g] * [e] = [g \circ e] = [g]$ and $[e] * [g] = [e \circ g] = [g]$. Finally, for any

$$[h] \in G/H, [h] * [h^{-1}] = [h \circ h^{-1}] = [e] = [h^{-1} \circ h] = [h^{-1}] * [h]$$

and so $[h]^{-1} = [h^{-1}]$.

Corollary 1. For any fixed $n \in \mathbb{N}$, $(Z_n, +)$ is an abelian group (Chapter Four, Section 5).

There are many other familiar mathematical objects that are groups. Any vector space, for instance, is an abelian group under addition (Appendix A). In particular then, the set of all 2×2 matrices with real entries is an abelian group under addition. The nonsingular 2×2 matrices are a group under multiplication, but this one is nonabelian. Another example of a nonabelian group that we have considered is the set of permutations on $\{1, 2, 3\}$ (Chapter Four, Exercises 7, problem 2). Here the binary operation is function composition, and because $a \circ b = \begin{pmatrix} 1 & 2 & 3 \\ 1 & 3 & 2 \end{pmatrix} = c$, whereas $b \circ a = \begin{pmatrix} 1 & 2 & 3 \\ 2 & 1 & 3 \end{pmatrix} = d$, we see that $a \circ b \neq b \circ a$.

Exercises 1

1. Let $(G, \circ)$ be a group and let H be a subgroup of G. Show that congruence modulo H is an equivalence relation on G. Do it in steps: (a) show that for any $g \in G, g \equiv g \bmod H$; (b) if $g, h \in G$ and $g \equiv h \bmod H$, show that $h \equiv g \bmod H$; (c) if g, h, k are in G and $g \equiv h$, $h \equiv k$, show that $g \equiv k$.

2. Consider the set $(Z_n, \cdot)$ for n fixed. Let E_n be the set of all elements of this set that have inverses. Show that $(E_n, \cdot)$ is an abelian group.

3. Let $(G, \circ)$ be an abelian group and let H be a subgroup of G. For each $g \in G$ let $\varphi_H(g) = [g] \in G/H$. So $\varphi_H : G \to G/H$.

    (a) Show that $\varphi_H(g \circ h) = \varphi_H(g) * \varphi_H(h)$ for all g, h in G.

    (b) What is $\{g \in G \mid \varphi_H(g) = [e]\}$?

2. Morphisms

We have had occasion to consider mappings $c : Z \to T$ such that $c(m + n) = c(m) \cdot c(n)$ for all $m, n \in Z$, and also mappings $c : Z_n \to T$ that have this property (Chapter Four, Section 6, Definition 2 and Definition 3). Mappings of this kind play a fundamental role in the general theory of groups.

Definition 1. Let $(G, \circ)$ and $(G', *)$ be two groups. A function $\varphi : G \to G'$ is said to be a homomorphism from G into G' if $\varphi(g \circ h) = \varphi(g) * \varphi(h)$ for all g, h in G.

In Chapter Four, Section 6 we defined certain maps that we called the characters of Z and of Z_n. We see now that these are simply the homomorphisms from $(Z, +)$ and $(Z_n, +)$, respectively, into $(T, \cdot)$.

The basic properties of any homomorphism are easily proved.

Lemma 1. Let $(G, \circ)$ and $(G', *)$ be two groups and let $\varphi : G \to G'$ be a homomorphism. Then (a) $\varphi(e) = e'$ here e is the identity element of G and e' is the identity of G'; (b) for every $g \in G$, $\varphi(g^{-1}) = \varphi(g)^{-1}$; (c) the set Ker $\varphi = \{g \in G \mid \varphi(g) = e'\}$ is a subgroup of G.

Proof. For any $g \in G$ we have $\varphi(g) = \varphi(g \circ e) = \varphi(g) * \varphi(e)$. Multiplying on the left by $\varphi(g)^{-1}$ we see that $e' = \varphi(e)$ as claimed. This proves (a). Now let $g \in G$ be given. We have $g^{-1} \circ g = g \circ g^{-1} = e$ and so $\varphi(g^{-1}) * \varphi(g) = \varphi(g) * \varphi(g^{-1}) = e'$ by (a). It now follows from Lemma 1 of the last section that $\varphi(g^{-1}) = \varphi(g)^{-1}$. This proves (b).

The set Ker φ is nonempty because, by (a), it contains e. To show that it is a subgroup we need only show that when g, h are in this set so also is

$g \circ h^{-1}$ (Section 1, Lemma 2). But Ker $\varphi(g) = \{g \in G \mid \varphi(g) = e'\}$, so all we must do is show that $\varphi(g \circ h^{-1}) = e'$ given that $\varphi(g) = e'$ and $\varphi(h) = e'$. Now $\varphi(g \circ h^{-1}) = \varphi(g) * \varphi(h^{-1}) = \varphi(g) * [\varphi(h)]^{-1} = e' * (e')^{-1} = e'$, here we have used (b).

Definition 2. Given a homomorphism $\varphi : G \to G'$, the set Ker $\varphi = \{g \in G \mid \varphi(g) = e'\}$, where e' is the identity element of G', is called the kernel of φ.

We have just shown that the kernel of any homomorphism on G is a subgroup of G.

Corollary 1. For $g, h \in G$ we have $\varphi(g) = \varphi(h)$ if, and only if, $g \circ h^{-1} \in$ Ker φ.

Proof. Suppose that $\varphi(g) = \varphi(h)$ then $\varphi(g) * \varphi(h)^{-1} = e'$, but because φ is a homomorphism, this can be written as $\varphi(g) * \varphi(h)^{-1} = \varphi(g) * \varphi(h^{-1}) = \varphi(g \circ h^{-1})$, and because this is equal to e', $g \circ h^{-1} \in$ Ker φ.

Now suppose that $g \circ h^{-1}$ is in the kernel of φ, then $e' = \varphi(g \circ h^{-1}) = \varphi(g) * \varphi(h^{-1}) = \varphi(g) * \varphi(h)^{-1}$. Multiplying on the right by $\varphi(h)$ we see that $\varphi(h) = \varphi(g)$.

Corollary 2. The homomorphism φ is one-to-one if, and only if, Ker $\varphi = \{e\}$.

Proof. Because we always have $\varphi(e) = e'$, assuming that φ is one-to-one immediately implies Ker $\varphi = \{e\}$. Conversely, if we assume Ker $\varphi = \{e\}$, then $\varphi(g) = \varphi(h)$ tells us that $g \circ h^{-1} = e$ and hence that $g = h$.

Definition 3. A homomorphism between two groups is called a monomorphism if it is one-to-one. A monomorphism that is also onto is called an isomorphism. Finally, two groups are said to be isomorphic if there is an isomorphism from one of these groups onto the other.

The set $\mathbb{R}$ is a group under addition and the set $\mathbb{R}_+$, the set of positive real numbers, is a group under multiplication. Note that $f(x) = e^x$ is an isomorphism from the group $(\mathbb{R}, +)$ onto the group $(\mathbb{R}_+, \cdot)$ and that $g(x) = \ln x$ is an isomorphism from $(\mathbb{R}_+, \cdot)$ onto $(\mathbb{R}, +)$. Long before the advent of the computer the fact that the log function is an isomorphism was exploited to simplify lengthy calculations. In these applications the base 10, the so-called common logarithms, was used.

Let $(G, \circ)$ be a group, and let K be a subgroup of G. Two elements g, h of G are congruent modulo K if $g \circ h^{-1}$ is in K, and this is an equivalence relation on G (Section 1, Lemma 3). We want to characterize the equivalence class defined by an element $g \in G$; i.e., the set $\{h \in G \mid g \equiv h \bmod K\}$ that we denoted by $[g]$. Note that $h \in [g]$ if, and only if, $h \circ g^{-1} \in K$. Thus $h \circ g^{-1} = k \in K$, and so $h = k \circ g$. This shows that $[g] \subseteq \{k \circ g \mid k \in K\}$; let's call this last set Kg. If $h \in Kg$, then $h = k \circ g$ for some $k \in K$ and hence $h \circ g^{-1} = k \in K$; i.e., $Kg \subseteq [g]$. So we have, for any $g \in G$, $[g] = Kg$. We will use this very soon. Note that $[g] * [h] = (Kg) * (Kh) = K(g \circ h)$.

Theorem 1. Let $(G, \circ)$ and $(G', *)$ be two *abelian* groups and let $\varphi : G \to G'$ be a homomorphism from G onto G'. Let $K = \text{Ker}\,\varphi$. Then the groups $(G/K, *)$ and $(G', *)$ are isomorphic. Furthermore, the map $\psi : G/K \to G'$ defined by $\psi(Kg) = \varphi(g)$ is an isomorphism between these two groups.

Proof. We first show that ψ is well defined. Suppose that $Kg = Kg'$ for g, g' in G, we must show that $\psi(Kg) = \varphi(g) = \varphi(g') = \psi(Kg')$. But because $Kg = Kg'$ and $e \in K$ (e is the identity of G), we see that $g = k \circ g'$ for some $k \in K$. Hence $\varphi(g) = \varphi(k \circ g') = \varphi(k) * \varphi(g') = e' * \varphi(g') = \varphi(g')$, where e' is the identity of G', and we have used the fact that $k \in K = \text{Ker } \varphi$. Thus ψ is well defined.

Given $g' \in G'$ there is a $g \in G$ such that $\varphi(g) = g'$, because φ is onto. But then $\psi(Kg) = g'$ showing that ψ is onto.

Suppose that $\psi(Kg) = \psi(Kg')$. Then, as we saw in the first paragraph of this proof, $\varphi(g) = \varphi(g')$ and so $\varphi(g) * \varphi(g')^{-1} = e'$. This says $g \circ (g')^{-1} \in \text{Ker } \varphi = K$. Hence $Kg = Kg'$, and we have shown that ψ is one-to-one.

Finally, $\psi[(Kg) * (Kg')] = \psi[K(g \circ g')] = \varphi(g \circ g') = \varphi(g) * (g') = \psi(Kg) * \psi(Kg')$.

Corollary 1. For any fixed $n \in \mathbb{N}$, the groups $(Z_n, +)$ and $(U_n, \cdot)$ are isomorphic.

Proof. Let ω be a primitive nth root of unity. Define $\varphi : Z \to U_n$ by setting $\varphi(k) = \omega^k$ for every $k \in Z$. Clearly, φ is a homomorphism because $\varphi(k + l) = \omega^{k+l} = \omega^k \omega^l = \varphi(k) \cdot \varphi(l)$. It is onto because ω is primitive. Note that $\text{Ker } \varphi = \{k \in Z \mid \varphi(k) = 1\} = \{k \in Z \mid \omega^k = 1\} = \{k \in Z \mid k = nl \text{ for some } l \in Z\} = \{nl \mid l \in Z\} = (n)$. Thus $Z/(n)$, which is $(Z_n, +)$, is isomorphic to $(U_n, \cdot)$.

Remark 1. Theorem 1 is also true for nonabelian groups. This is because the kernel of a homomorphism is a special kind of subgroup, called a normal subgroup. We shall not discuss these matters any further (but see Exercises 2, problem 4) except to say that when G is nonabelian, G/H need not be a group.

Exercises 2

1. Let $(G, \circ)$ and $(G^*, *)$ be two groups, and let $\varphi : G \to G^*$ be a homomorphism. Show that $\text{Im } \varphi = \{\varphi(g) \mid g \in G\}$ is a subgroup of $(G^*, *)$.

2. Let $(G, \circ)$ be a group, and let $\varphi : G \to G$ be an isomorphism. Because φ maps G onto itself, we call φ an automorphism.

   (a) Show that $H = \{g \in G \mid \varphi(g) = g\}$ is a subgroup of G.

   (b) Define $\varphi : (\mathbb{C}, +) \to (\mathbb{C}, +)$ by setting $\varphi(z) = \bar{z}$ (the complex conjugate of z) for each $z \in \mathbb{C}$. Show that this is an automorphism and identify $\{z \in \mathbb{C} \mid \varphi(z) = z\}$.

3. Show that $f(x) = e^{ix}$ is a monomorphism from $(\mathbb{R}, +)$ into $(\mathbb{C} \setminus \{\circ\}, \cdot)$ and identify $\text{Im}(f)$.

4. Given a group $(G, \circ)$ we may define, for each fixed $g \in G$, $\varphi_g(h) = g^{-1} \circ h \circ g$ for all $h \in G$.

   (a) Show that $\varphi_g : G \to G$ is an automorphism (problem 2 above).

   (b) A subgroup H of G is said to be a normal subgroup if $\varphi_g(H) = \{\varphi_g(h) \mid h \in H\} \subseteq H$ for every fixed $g \in G$. Show that the kernel of any homomorphism on G is a normal subgroup.

   (c) If $(G, \circ)$ is abelian, show that every subgroup of G is a normal subgroup.

3. Rings

The subgroups of the group $(Z, +)$ are of the form $(l) = \{kl \mid k \in Z\}$ where $l = 0, 1, 2, \ldots$ is fixed. These have the unusual property that $n \cdot a \in (l)$ whenever $a \in (l)$ and $n \in Z$. This is what enabled us to define multiplication in Z_l (Chapter Four, Section 8, Definition 3, Lemma 4 and the discussion just before Lemma 4). We can get a better understanding of what is happening here by introducing another abstract algebraic structure.

Definition 1. A ring $(R, +, \cdot)$ is a nonempty set R that is closed under two associative binary operations $+$ and $\cdot$ for which (a) $(R, +)$ is an abelian group; (b) given a, b, c in R, we have both $a \cdot (b+c) = a \cdot b + a \cdot c$ and $(a+b) \cdot c = a \cdot c + b \cdot c$. If we also have (c) $a \cdot b = b \cdot a$ for all a, b in R, then we may say that the ring $(R, +, \cdot)$ is commutative.

Given a ring $(R, +, \cdot)$ the identity element of $(R, +)$ is denoted by 0 and the inverse of any $a \in R$ is denoted by $(-a)$. When R has had a multiplicative identity, we shall denote it by 1 and say that $(R, +, \cdot)$ is a ring with unit. To avoid triviality, we shall always assume, given a ring with unit, that $0 \neq 1$.

It is clear that $(Z, +, \cdot)$ is a commutative ring with unit. The set of all 2×2 matrices with real entries is, under matrix addition and matrix multiplication, a noncommutative ring that has a unit. Let us show that the laws of signs, that we all memorized in high school, are consequences of the axioms for a ring.

Lemma 1. In any ring $(R, +, \cdot)$ we have (i) $a \cdot 0 = 0 \cdot a = 0$ for every $a \in R$; (ii) $a \cdot (-b) = -(a \cdot b) = (-a) \cdot b$ for all a, b in R; (iii) $(-a) \cdot (-b) = a \cdot b$ for all a, b in R.

Proof. For any a, b in R we have $a \cdot b = a \cdot (b+0) = a \cdot b + a \cdot 0$. Thus $a \cdot 0 = 0$. Similarly, $b \cdot a = (b+0) \cdot a = b \cdot a + 0 \cdot a$ and so $0 \cdot a = 0$. Now that (i) has been proved we may use it to prove (ii) and (iii). Because $0 = a \cdot 0 = a \cdot [b + (-b)] = a \cdot b + a \cdot (-b)$ we see that $a \cdot (-b)$ is the inverse, in $(R, +)$, of $a \cdot b$. But there is only one inverse of $a \cdot b$, $-(a \cdot b)$ (Section 1, Lemma 1). Thus $a \cdot (-b) = -(a \cdot b)$. The proof that $(-a) \cdot b = -(a \cdot b)$ is similar.

Finally, $0 = (-a) \cdot 0 = (-a) \cdot [b + (-b)] = (-a) \cdot b + (-a) \cdot (-b)$. Using (ii) we may rewrite this as $0 = -(a \cdot b) + (-a) \cdot (-b)$. But this clearly shows that $(-a) \cdot (-b)$ is the inverse in $(R, +)$ of $-(a \cdot b)$ and since that inverse is $a \cdot b$, $(-a) \cdot (-b) = a \cdot b$.

In the study of a given ring, certain of its subrings are of special importance. We define these now.

Definition 2. Let $(R, +, \cdot)$ be a ring. A nonempty subset I of R is called an ideal in R if (a) I is a subgroup of $(R, +)$; (b) for any $a \in I$ and any $r \in R$ both $a \cdot r$ and $r \cdot a$ are in I. An ideal I is called a proper ideal if $I \neq R$.

Remark 1. We have seen that $(l^1(Z), +, *)$ is a commutative ring with unit (Chapter Four, Section 1, Definition 1, Lemma 3 and Exercises 1, problem 2). When we defined the ideals in this ring, however, we were looking at it as an "algebra over $\mathbb{C}$" (Chapter Four, Section 2, Definition 2). A ring $(R, +, \cdot)$ is called an "algebra over $\mathbb{C}$" if (a) $(R, +)$ is a vector space over $\mathbb{C}$ (Appendix A); (b) for each a, b in R and each $\lambda \in \mathbb{C}$, $\lambda(a \cdot b) = (\lambda a) \cdot b = a \cdot (\lambda b)$. In a ring $(R, +, \cdot)$ we want our ideals to be subgroups of $(R, +)$; in an algebra $(R, +, \cdot)$ over $\mathbb{C}$ we want our ideals to be linear subspaces of the vector space $(R, +)$. The proof that $(l^1(Z), +, *)$ is an algebra over $\mathbb{C}$ was given in (Chapter Three, Section 1, example (iv) and Chapter Four, Exercises 1, problem 2(b)).

Let us focus now on a fixed commutative ring with unit, $(R, +, \cdot)$, and a proper ideal I in R. Because I is a subgroup of the abelian group $(R, +)$, defining $a \equiv b \bmod I$ to mean $(a - b) \in I$ gives us an equivalence relation on R (Section 1, Lemma 3). Also, the equivalence classes are the sets $a + I = \{a + b \mid b \in I\}$ (Section 2, the discussion just above Theorem 1). We know that R/I is an abelian group when we define $(a + I) + (b + I)$ to be $(a + b) + I$. We can define a multiplication in R/I. What we do is set $(a + I) * (b + I)$ equal to $a \cdot b + I$. We shall now prove:

Theorem 1. $(R/I, +, *)$ is a commutative ring with unit. The map $\varphi : R \to R/I$ defined by setting $\varphi(a) = a + I$ for all $a \in R$ satisfies (i) $\varphi(a+b) = \varphi(a) + \varphi(b)$ for all a, b in R; (ii) $\varphi(a \cdot b) = \varphi(a) * \varphi(b)$ for all a, b in R.

Proof. We already know that $(R/I, +)$ is an abelian group and that the map φ satisfies (a). Let us show that $*$ is well defined. To do this we must show that if $a \equiv a' \bmod I$ and $b \equiv b' \bmod I$ then $a \cdot b \equiv a' \cdot b' \bmod I$. Now $(a - a') \in I$, hence $(a - a') \cdot b \in I$. Also, because $b - b'$ is in I, $a' \cdot (b - b') \in I$. Thus we have both $(a - a') \cdot b = a \cdot b - a' \cdot b$ in I and $a' \cdot (b - b') = a' \cdot b - a' \cdot b'$ in I. Adding we get $a \cdot b - a' \cdot b'$ in I which says $a \cdot b \equiv a' \cdot b' \bmod I$, and this is what we wanted to show.

Observe that $*$ is commutative for $(a + I) * (b + I) = a \cdot b + I = b \cdot a + I = (b + I) * (a + I)$ because $(R, \cdot)$ is commutative. Similarly, $*$ is associative. Note that $(a + I) * [(b + I) + (c + I)] = (a + I) * [(b + c) + I] = a \cdot (b + c) + I = (a \cdot b + a \cdot c) + I = (a \cdot b + I) + (a \cdot c + I)$, showing that $(R/I, +, *)$ is a ring. The fact that $(1 + I)$ is a unit in this ring and that φ satisfies (ii) are immediate from the definitions.

Corollary 1. For each fixed l, $(Z_l, +, \cdot)$ is a commutative ring with unit.

Proof. $Z_l = Z/(l)$ and the set $(l) = \{kl \mid k \in Z\}$ is an ideal in $(Z, +, \cdot)$. Because this last ring is commutative and has a unit, the corollary is an immediate consequence of our theorem.

Exercises 3

1. We have seen that the ideals in $(Z, +, \cdot)$ are of the form $(a) = \{k \cdot a \mid k \in Z\}$.

    (a) Show that $(a) \subseteq (b)$ if, and only if, $b \mid a$.

    (b) A proper ideal I in a ring $(R, +, \cdot)$ is said to be a maximal ideal if the only ideals that contain I are R and I itself. What are the maximal ideals in $(Z, +, \cdot)$?

2. We refer to Chapter Four, Section 9 and the discussion just before Lemma 2. Let q be a fixed odd prime number and let $X = \{a + b\sqrt{3} \mid a, b \in Z_q\}$. For $a_1 + b_1\sqrt{3}$ and $a_2 + b_2\sqrt{3}$ in X define their sum to be $(a_1 + a_2) + (b_1 + b_2)\sqrt{3}$, where $a_1 + a_2$ and $b_1 + b_2$ are computed in Z_q.

    (a) Show that $(X, +)$ is an abelian group.

    (b) Define $(a_1 + b_1\sqrt{3}) \cdot (a_2 + b_2\sqrt{3})$ to be $(a_1 a_2 + 3b_1 b_2) + (a_1 b_2 + a_2 b_1)\sqrt{3}$, where $a_1 a_2 + 3b_1 b_2$ and $a_1 b_2 + a_2 b_1$ are computed in Z_q. Show that $(X, +, \cdot)$ is a commutative ring with unit.

3. Let $(R, +, \cdot)$ be a commutative ring with unit. For any fixed $a \in R$ let $(a) = \{r \cdot a \mid r \in R\}$. Show that (a) is an ideal in R.

4. Show that, in a commutative ring with unit, every proper ideal is contained in a maximal ideal (defined in 1(c) above).

4. Fields

We are all familiar with $(Q, +, \cdot)$, $(R, +, \cdot)$, and $(\mathbb{C}, +, \cdot)$. Each of these is a commutative ring with unit, but they are very special rings in that each of their nonzero elements has a multiplicative inverse. Such rings are called fields, and in this section we discuss finite fields, sometimes called Galois fields. These are of interest in pure mathematics, and they have proved useful in digital signal processing [see Note 2].

Definition 1. A field $(F, +, \cdot)$ is a commutative ring with unit in which every nonzero element has a multiplicative inverse.

We have seen that $(Z_l, +, \cdot)$ is a commutative ring with unit (Section 3, Corollary 1 to Theorem 1). Furthermore, by Corollary 1 to Theorem 3 (Chapter Four, Section 8), $(Z_l, +, \cdot)$ is a field if, and only if, l is a prime.

Given a field $(F, +, \cdot)$ it is convenient to denote the additive and multiplicative identities of F by $\bar{0}$ and $\bar{1}$, respectively. If $a \in F$, we set $1 \cdot a = a$, $2 \cdot a = a + a$, and in general $n \cdot a = a + (n-1) \cdot a$, $n = 3, 4, \ldots$. Consider now the sequence of field elements $1 \cdot \bar{1}, 2 \cdot \bar{1}, 3 \cdot \bar{1}, \ldots$ There are two possibilities: We can have $m \cdot \bar{1} \neq n \cdot \bar{1}$ whenever $m \neq n$, or we can have $n \cdot \bar{1} = m \cdot \bar{1}$ for some m, n with $n < m$. In the latter case we see that $(m - n) \cdot \bar{1} = \bar{0}$.

Definition 2. Let $(F, +, \cdot)$ be a field. If $n \cdot \bar{1} \neq \bar{0}$ for all $n \in \mathbb{N}$, we shall say that F has characteristic zero. If F does not have characteristic zero, then the smallest element n of $\mathbb{N}$ such that $n \cdot \bar{1} = \bar{0}$ is taken to be the characteristic of F.

It is clear that Q, R, and $\mathbb{C}$ each have characteristic zero and that Z_p has characteristic p.

Theorem 1. The characteristic of a field is either zero or a prime number.

Proof. If $(F, +, \cdot)$ is a field having characteristic $n > 0$, then n is the smallest positive integer such that $n \cdot \bar{1} = \bar{0}$. If n is a composite number, then $n = pq$ where $1 < p < n$ and $1 < q < n$. It is clear, however, that $\bar{0} = n \cdot \bar{1} = (pq) \cdot \bar{1} = (p \cdot \bar{1}) \cdot (q \cdot \bar{1})$. If the field element $p \cdot \bar{1} \neq \bar{0}$, it has a multiplicative inverse $(p \cdot \bar{1})^{-1}$. Hence, recalling that $(p \cdot \bar{1})^{-1} \cdot \bar{0} = \bar{0}$ (Section 3, Lemma 1, (a)), we see that $\bar{0} = (p \cdot \bar{1})^{-1} \cdot \bar{0} = (p \cdot \bar{1})^{-1} \cdot [(p \cdot \bar{1})(q \cdot \bar{1})] = q \cdot \bar{1}$. But this is impossible because $1 < q < n$ and n is the characteristic of F.

Now let $(F, +, \cdot)$ be a field with characteristic $p \neq 0$. Set $0 \cdot \bar{1} = \bar{0}$ and note that $\{0 \cdot \bar{1}, 1 \cdot \bar{1}, 2 \cdot \bar{1}, \ldots\} = \{(n-1) \cdot \bar{1} \mid n \in \mathbb{N}\}$ is just the set $\{0 \cdot \bar{1}, 1 \cdot \bar{1}, \ldots, (p-1)\bar{1}\}$. Furthermore, if $m \cdot \bar{1}$ and $n \cdot \bar{1}$ are in this last set, then $m \cdot \bar{1} + n \cdot \bar{1} = (m + n) \cdot \bar{1}$ and $(m \cdot \bar{1}) \cdot (n \cdot \bar{1}) = (mn) \cdot \bar{1}$ provided that the sum, $m + n$, and the product, mn, are carried out mod p.

Now we can define a map $\varphi : Z_p \to F$ by setting $\varphi(n) = n \cdot \bar{1}$ for each $n \in Z_p$. We are using n to mean two different things. In $\varphi(n)$, n is an element of Z_p, and in $n \cdot \bar{1}$, n is the integer n. This should not cause the reader any trouble. Introducing new notation to distinguish the two n's seems unnecessarily pedantic at this point.

Observe that $\varphi(m+n) = \varphi(m) + \varphi(n)$ and $\varphi(mn) = \varphi(m) \cdot \varphi(n)$ for all m, n in Z_p. We have shown that every field of characteristic p "contains" the field Z_p.

Theorem 2. Let $(F, +, \cdot)$ be a field with a finite, say q, number of elements. Then $q = p^n$, where p is the characteristic of F, and $n \in \mathbb{N}$.

Proof. It is clear, because F has only finitely many elements, that its characteristic must be $p \neq 0$. Now F contains Z_p, and we may regard F to be a vector space whose scalars come from Z_p (Appendix A.) Let $v_1, ..., v_n$ be a basis for this vector space. Then every element $x \in F$ can be written in one, and only one, way as a linear combination of the basis elements. Hence $x = \sum_{j=1}^{n} \alpha_j v_j$, where each $\alpha_j \in Z_p$. Note also that any such sum is in F. Because there are p choices for each coefficient and there are n coefficients, F contains p^n elements.

Remark 1. It is true, although we shall not prove it, that for any prime p and any $n \subset \mathbb{N}$ there is a field containing p^n elements. Furthermore, there is only one such field in that any two are isomorphic. As we mentioned above, such fields are called Galois fields. They are often denoted by $GF(p^n)$.

Exercises 4

1. Let $(F, +, \cdot)$ be a field and let $a, b \in F$. If $a \cdot b = \bar{0}$, show that either $a = \bar{0}$ or $b = \bar{0}$.

    (a) If $(F, +, \cdot)$ is a field, show that the only ideals in F are $\{\bar{0}\}$ and F itself.

    (b) If $(R, +, \cdot)$ is a commutative ring with unit and if the only ideals in R are $\{0\}$ and R itself, show that $(R, +, \cdot)$ is a field

2. Let $(F, +, \cdot)$ be a field with characteristic $p \neq 0$.

    (a) For any $a, b \in F$, show that $(a+b)^p = a^p + b^p$. Hint: Use the binomial theorem and the fact that p is a prime.

    (b) Let $G = \{x^p \mid x \in F\}$. Show that $(G, +, \cdot)$ is a field.

Notes

1. A very accessible introduction to modern algebra can be found in "Abstract Algebra" by J. N. Herstein (Macmillan Publishing Co., 2nd Edition, New York, 1990).

2. Groups, rings, fields, and their application to digital signal processing are discussed in "Fast Algorithms for Digital Signal Processing" by Richard E. Blahut (Addison-Wesley Publishing Co., Reading, MA, 1985). See pp. 21–33. The Fourier transform in an arbitrary field is discussed on p. 31. There is also a treatment of finite fields (pp. 155–165). Euler's phi function is discussed on pp. 156 and 157, and Mersenne numbers and associated finite fields are used on pp. 182–185.

3. There are many applications of group theory to various areas of physics. These are discussed in "Applications of Finite Groups" by John S. Lomont (Academic Press, New York, 1961).

Appendix A

Linear Algebra

Here we shall recall some terminology from linear algebra

Definition 1. A nonempty set V is said to be a vector space over $\mathbb{K}$ if it has all of the following properties:

1. There is a rule that assigns to each pair u, v of V a unique element of V called their sum and denoted by $u + v$. This rule is called vector addition.

2. Vector addition satisfies

    (a) The Commutative Law; i.e., $u + v = v + u$ for all u, v in V.

    (b) The Associative Law; i.e., $(u + v) + w = u + (v + w)$ for all u, v, w in V.

    (c) There is a unique element $0 \in V$ for which $u + 0 = 0 + u$ for all $u \in V$. We call 0 the additive identity of V.

    (d) For each $u \in V$ there is a unique element $(-u) \in V$ such that $u + (-u) = (-u) + u = 0$.

3. There is a rule that assigns to each pair α, v, where $\alpha \in \mathbb{K}$ and $v \in V$, a unique element of V called the product of α and v denoted by αv. The rule is called scalar multiplication.

4. Scalar multiplication satisfies.

    (a) $\alpha(\beta u) = (\alpha \beta) u$ for all α, β in $\mathbb{K}$ and all $u \in V$;

    (b) $\alpha(u + v) = \alpha u + \alpha v$ for all α in $\mathbb{K}$ and all $u, v \in V$;

(c) $(\alpha + \beta)u = \alpha v + \beta u$ for all $\alpha\beta$ in $\mathbb{K}$ and all $u \in V$;

(d) $1u = u$ for all $u \in V$.

The elements of a vector space V over $\mathbb{K}$ are called vectors, the elements of $\mathbb{K}$ are called scalars, and $\mathbb{K}$ is called the field of scalars.

Now let V be a vector space over $\mathbb{K}$.

Definition 2. A nonempty subset W of V is said to be a linear subspace if

(a) For any two vectors u, v, in W the vector $u + v$ is in W;

(a) For any vector w in W and any scalar α (i.e., any element of α of $\mathbb{K}$), the vector αw is in W.

Let S be a nonempty subset of V. A vector v is said to be a linear combination of vectors in S if there is a finite set $u_1, ...u_k$ in S and scalars $\alpha_1, ...\alpha_k$ such that $v = \sum_{j=1}^{k} a_j u_j$. The set of all vectors in V that are linear combinations of vectors in S is denoted by lin S. It is easy to see that lin S is a linear subspace that contains S. We call it the linear subspace spanned by, or generated by, S. In particular, if lin $S = V$ we say that S spans V.

Definition 3. Let V, W be two vector spaces over the *same* $\mathbb{K}$, and let $T : V \to W$. We shall say that T is a linear function, or linear map, if (i) $T(u + v) = T(u) + T(v)$ for all u, v in V; (ii) $T(\alpha v) = \alpha T(v)$ for all α in $\mathbb{K}$ and all v in V.

Corresponding to any linear map $T : V \to W$ we have two linear subspaces: the range of T (Definition 1 of Section 2, Chapter Zero), which is a linear subspace of W; and the null space, or kernel, of T defined to be $\{v \in V \mid T(v) = 0\}$.

Appendix B
The Completion

Definition 1. Let $(E, \|\cdot\|)$ be a normed space. A Banach space $(\bar{E}, \|\cdot\|_1)$ is said to be a completion of $(E, \|\cdot\|)$ if

(a) There is a linear map $T : E \to \bar{E}$ such that $\|T\vec{x}\|_1 = \|\vec{x}\|$ for every $\vec{x} \in E$ (we say that the map T is norm preserving);

(b) The linear subspace $T(E) = \{T\vec{x} \mid \vec{x} \in E\}$ is dense in $(\bar{E}, \|\cdot\|_1)$.

We shall show that every normed space has a completion and that, in a certain sense, this completion is unique. The construction is quite straightforward; it is similar to Cantor's construction of the reals from the rationals but lengthy and a little bit tedious. We shall leave some of the more routine verifications to the reader so as not to obscure the underlying ideas. We begin with some simple facts about Cauchy sequences (Definition 3 of Section 2, Chapter Three) in a normed space $(E, \|\cdot\|)$.

1. if $\vec{x}, \vec{y}$ are in E, then $\left| \|\vec{x}\| - \|\vec{y}\| \right| \leq \|\vec{x} - \vec{y}\|$.

2. If $\{\vec{x}_n\} \subseteq E$ is a Cauchy sequence, then the sequence of real numbers $\{\|\vec{x}_n\|\}$, is convergent.

3. If $\{\vec{x}_n\}$ and $\{\vec{y}_n\}$ are two Cauchy sequences in E and if $\lim_{n \to \infty} (\vec{x}_n - \vec{y}_n) = 0$, then $\lim_{n \to \infty} \|\vec{x}_n\| = \lim_{n \to \infty} \|\vec{y}_n\|$.

We read the proofs for the reader.

Appendix B ■ The Completion

Theorem 1. Every normed space has a completion.

Proof. Let $(E, \|\cdot\|)$ be a normed space and let C be the set of all Cauchy sequences of points of E. We may define an equivalence relation (Chapter Zero, Section 2) on C by agreeing to say $\{\vec{x}_n\}$ and $\{\vec{y}_n\}$ are equivalent when $\lim \|\vec{x}_n - \vec{y}_n\| = 0$. For each $\{\vec{x}_n\} \in C$, let $[\{\vec{x}_n\}]$ be the equivalence class (Chapter Zero, Exercises 2, Problem 1) that contains $\{\vec{x}_n\}$, and let $\bar{E}$ be the set of all equivalence classes in C. We can define two operations on the elements of $\bar{E}$ as follows: For any $[\{\vec{x}_n\}], [\{\vec{y}_n\}]$ we let $[\{\vec{x}_n\}] + [\{\vec{y}_n\}] = [\{\vec{x}_n + \vec{y}_n\}]$, and for any scalar λ we let $\lambda [\{\vec{x}_n\}] = [\{\lambda \vec{x}_n\}]$. It is easy to show that $\bar{E}$, with these operations, is a vector space.

We have seen that for any $\{\vec{x}_n\} \in C$, $\lim_{n \to \infty} \|\vec{x}_n\|$ exists ((2) above). Furthermore if $\{\vec{x}_n\}$ and $\{\vec{y}_n\}$ are equivalent, then $\lim_{n \to \infty} \|\vec{x}_n\| = \lim_{n \to \infty} \|\vec{y}_n\|$ ((3) above). Thus we may define, for any $[\{\vec{x}_n\}] \in \bar{E}$,

$$\left\| [\{\vec{x}_n\}] \right\|_1 = \lim_{n \to \infty} \|\vec{x}_n\|$$

It is easy to see that $\|\cdot\|_1$ is a norm on $\bar{E}$.

Now given any $\vec{x} \in E$ we consider the sequence $\vec{x}, \vec{x}, \ldots$; i.e., the element $\{\vec{x}_n\} \in C$ such that $\vec{x}_n = \vec{x}$ all n. Then, denoting this sequence by $\{\vec{x}\}$, $[\{\vec{x}\}] \in \bar{E}$ and $\left\| [\{\vec{x}\}] \right\|_1 = \|\vec{x}\|$. We define a map $T : E \to \bar{E}$ by setting $T\vec{x} = [\{\vec{x}\}]$ for every $\vec{x} \in E$. It is clear that T is a well-defined linear map that is norm preserving (i.e., $\|T\vec{x}\|_1 = \|\vec{x}\|$ for all $\vec{x}$).

Let us show that the linear subspace $T(E)$ is dense in $(\bar{E}, \|\cdot\|_1)$. Given $[\{\vec{x}_n\}] \in \bar{E}$ the sequence $\{\vec{x}_n\}$ is in C and so, given $\varepsilon > 0$, there is an N such that

$$\|\vec{x}_n - \vec{x}_m\| < \varepsilon$$

whenever $m, n \geq N$. Choose, and fix $k \geq N$ and consider the sequence $\vec{x}_k, \vec{x}_k, \ldots$. We have

$$[\{\vec{x}_n\}] - [\{\vec{x}_k, \vec{x}_k, \ldots\}] = [\{\vec{x}_n - \vec{x}_k\}_{n=1}^{\infty}]$$

and

$$\left\| \left[\{\vec{x}_n\} \right] - \left[\{\vec{x}_k, \vec{x}_k, \ldots\} \right] \right\|_1 = \left\| \left[\{\vec{x}_n - \vec{x}_k\}_{n=1}^{\infty} \right] \right\|_1$$
$$= \lim_{n \to \infty} \left\| \vec{x}_n - \vec{x}_k \right\| < \varepsilon$$

Thus, as $k \to \infty$, the sequence

$$\left[\{\vec{x}_1, \vec{x}_1, \ldots\} \right], \left[\{\vec{x}_2, \vec{x}_2, \ldots\} \right], \ldots \left[\{\vec{x}_k, \vec{x}_k, \ldots\} \right], \ldots$$

converges to the given element of $\bar{E}$ for $\|\cdot\|_1$. Because each term of this sequence is in $T(E)$, we see that $T(E)$ is dense in $(\bar{E}, \|\cdot\|_1)$ as claimed.

Note that we have just shown that for any $\left[\{\vec{x}_n\} \right]$ in $\bar{E}$ the sequence $T\vec{x}_1, T\vec{x}_2, \ldots T\vec{x}_k, \ldots$ converges to $\left[\{\vec{x}_n\} \right]$ for $\|\cdot\|_1$. We use this observation below.

To finish the proof we need only show that $(\bar{E}_1 \|\cdot\|_1)$ is complete. So let $\{\vec{V}_n\}$ be a Cauchy sequence in $\bar{E}$ and choose, as we may because $T(E)$ is dense in $\bar{E}$, $\vec{v}_n \in E$ such that

$$\left\| T\vec{v}_n - \vec{V}_n \right\|_1 < \frac{1}{n}$$

We do this for each $n = 1, 2, \ldots$. Consider the sequence $\{\vec{v}_n\}$ contained in E that we have constructed. We note that

$$\left\| \vec{v}_n - \vec{v}_m \right\| = \left\| T\left(\vec{v}_n - \vec{v}_m \right) \right\|_1 = \left\| T\vec{v}_n - T\vec{v}_m \right\|_1$$
$$\leq \left\| T\vec{v}_n - \vec{V}_n \right\|_1 + \left\| \vec{V}_n - \vec{V}_m \right\|_1 + \left\| \vec{V}_m - T\vec{v}_m \right\|_1$$
$$\leq \frac{1}{n} + \left\| \vec{V}_n - \vec{V}_m \right\|_1 + \frac{1}{m}$$

where we have used the fact that T is norm preserving. Because $\{\vec{V}_n\}$ is a Cauchy sequence, our inequality shows that $\{\vec{v}_n\}$ is also a Cauchy sequence. Set $\vec{V} = \left[\{\vec{v}_n\} \right]$ and recall that $T\vec{v}_1, T\vec{v}_2, \ldots T\vec{v}_k, \ldots$ converges to $\vec{V}$ for $\|\cdot\|_1$.

It follows from the inequality

$$\left\|\vec{V}_n - \vec{V}\right\|_1 \le \left\|\vec{V}_n - T\vec{v}_n\right\|_1 + \left\|T\vec{v}_n - \vec{V}\right\|_1 \le \frac{1}{n} + \left\|T\vec{v}_n - \vec{V}\right\|_1$$

that the Cauchy sequence $\left\{\vec{V}_n\right\}$ is convergent to a point of $\bar{E}$.

While we have the proof before us, let us show that when $(V, \langle, \rangle)$ is an inner product space and $\|\cdot\|$ is the norm induced on V by the inner product, then the construction applied to $(V, \|\cdot\|)$ gives us a space $(\bar{V}, \|\cdot\|)$ that is actually a Hilbert space.

First note that when $\{\vec{v}_n\}, \{\vec{w}_n\}$ are in C (the set of all Cauchy sequences of $(V, \|\cdot\|)$) then

$$\left|\langle \vec{v}_n, \vec{w}_n \rangle - \langle \vec{v}_m, \vec{w}_m \rangle\right| =$$
$$\left|\langle \vec{v}_n - \vec{v}_m, \vec{w}_n - \vec{w}_m \rangle\right| \le \left\|\vec{v}_n - \vec{v}_m\right\| \left\|\vec{w}_n - \vec{w}_m\right\|$$

by Theorem 2 of Section 3, Chapter 3. Hence $\lim_{n \to \infty} \langle \vec{v}_n, \vec{w}_n \rangle$ exists in $\mathbb{K}$. Furthermore, if $\{\vec{v}'_n\}$ and $\{\vec{w}'_n\}$ are equivalent, respectively, to $\{\vec{v}_n\}$ and $\{\vec{w}_n\}$, then

$$\left|\langle \vec{v}_n, \vec{w}_n \rangle - \langle \vec{v}'_n, \vec{w}'_n \rangle\right| \le \left\|\vec{v}_n - \vec{v}'_n\right\| \left\|\vec{w}_n - \vec{w}'_n\right\|$$

showing that $\lim \langle \vec{v}_n, \vec{w}_n \rangle = \lim \langle \vec{v}'_n, \vec{w}'_n \rangle$. Hence we may define, for any $[\{\vec{v}_n\}], [\{\vec{w}_n\}]$ in $\bar{V}$, an inner product by setting

$$\langle [\{\vec{v}_n\}], [\{\vec{w}_n\}] \rangle = \lim \langle \vec{v}_n, \vec{w}_n \rangle$$

It is easy to see that the norm associated with this inner product is just $\|\cdot\|_1$. Thus $(\bar{V}, \|\cdot\|_1)$ is complete showing that $(\bar{V}, \langle, \rangle)$ is a Hilbert space.

Finally, we note that the map $T: V \to \vec{V}$ defined in the proof of Theorem 1 satisfies

$$\langle T\vec{u}, T\vec{v} \rangle = \langle \vec{u}, \vec{v} \rangle$$

for all $\vec{u}, \vec{v}$ in V.

We have proved:

Theorem 2. Given any inner product space $(V, \langle,\rangle)$ there is a Hilbert space $(\bar{V}, \langle,\rangle)$ and there is a linear map $T : V \to \bar{V}$ such that $\langle T\vec{u}, T\vec{v} \rangle = \langle \vec{u}, \vec{v} \rangle$ for all $\vec{u}, \vec{v}$ in V (we say that T preserves the inner product), and the linear subspace $T(V)$ is dense in $\bar{V}$.

Definition 2. Given an inner product space $(V, \langle,\rangle)$ a Hilbert space having the properties stated in Theorem 2 is called a completion of $(V, \langle,\rangle)$.

Let us now show that all completions of a normed or an inner product space are "essentially" the same.

Theorem 3. Given any two completions of a normed space, there is a linear norm preserving map from either of these completions onto the other. Given any two completions of an inner product space, there is a linear map from either of these completions onto the other that preserves inner product.

Proof. Let $(E, \|\cdot\|)$ be a normed space, and let $(E_1, \|\cdot\|_1)$ and $(E_2, \|\cdot\|_2)$ be two completions of this space. Then we have two linear norm-preserving maps $T_1 : E \to E_1$ and $T_2 : E \to E_2$ such that $T_1(E)$ is dense in $(E_1, \|\cdot\|_1)$ and $T_2(E)$ is dense in $(E_2, \|\cdot\|_2)$. To simplify the notation, let us simply say that E is a dense linear subspace of both $(E_1, \|\cdot\|_1)$ and $(E_2, \|\cdot\|_2)$.

We may define a map $T : E_1 \to E_2$ as follows: For each $\vec{x} \in E \subseteq E_1$ we set $T(\vec{x}) = \vec{x} \in E_2$. For $\vec{y} \in E_1 \setminus E$ we find first a sequence $\{\vec{x}_n\} \subseteq E$ such that $\lim \vec{x}_n = \vec{y}$ for $\|\cdot\|_1$. Then $\{\vec{x}_n\}$ is a Cauchy sequence for $\|\cdot\|_1$ and $\{T(\vec{x}_n) = \vec{x}_n\}$ is a Cauchy sequence for $\|\cdot\|_2$. We set $T(\vec{y})$ equal to the limit of $\{T(\vec{x}_n)\}$ for $\|\cdot\|_2$.

Note that if $\{\vec{x}_n\}$ and $\{\vec{x}'_n\}$ both converge to $\vec{y}$ for $\|\cdot\|_1$, $\{\vec{x}_n - \vec{x}'_n\}$ converges to zero for both $\|\cdot\|_1$ and $\|\cdot\|_2$; because $\{\vec{x}_n\}$ and $\{\vec{x}'_n\}$ are in E. Thus $\lim T(\vec{x}_n) = \lim T(\vec{x}'_n)$ (for $\|\cdot\|_2$), and we see that T is well defined.

It is easy to see that T is linear. Hence to show that T is one-to-one (Chapter Zero, Section 2) we need only show that $\{\vec{y} \in E_1 \mid T(\vec{y}) = \vec{0} \in E_2\}$ contains only the zero vector of E_1. Suppose that $\vec{y}$ is in this last set, and let $\{\vec{x}_n\}$ be a sequence in E that converges to $\vec{y}$ for $\|\cdot\|_1$. But then $\lim \vec{x}_n = \lim T(\vec{x}_n) = \vec{0} \in E_2$ for $\|\cdot\|_2$ and so, because $\|\cdot\|_1 = \|\cdot\|_2$ on E, $\lim \vec{x}_n = \vec{0} \in E_1$ for $\|\cdot\|_1$. Thus $\vec{y}$ is the zero vector of E_1.

Given any $\vec{z} \in E_2$ we can find a sequence $\{\vec{x}_n\} \subseteq E$ that converges to $\vec{z}$ for $\|\cdot\|_2$. But then $\{\vec{x}_n\}$ converges to an element $\vec{y} \in E_1$ for $\|\cdot\|_1$; because it is Cauchy for E_1. Clearly then, $T(\vec{y}) = \vec{z}$, and we have shown that the map T is onto.

Finally, let $\vec{y} \in E_1$, and consider $\|\vec{y}\|_1$ and $\|T(\vec{y})\|_2$. We have $T(\vec{y}) = \lim T(\vec{x}_n)$ where $\{\vec{x}_n\} \subseteq E$ and $\lim \vec{x}_n = \vec{y}$ for $\|\cdot\|_1$. But then $\lim \|\vec{x}_n\|_1 = \|\vec{y}\|_1$ and $\lim \|\vec{x}_n\|_2 = \lim \|T(\vec{x}_n)\|_2 = \|T(\vec{y})\|_2$ (Chapter Three, Exercises 2, problem 1(a)). Because $\|\cdot\|_1$ and $\|\cdot\|_2$ coincide on E, we see that $\|T(\vec{y})\|_2 = \|\vec{y}\|_1$ and we have proved that T is norm preserving.

Now let $(V, \langle,\rangle)$ be an inner product space, and let $(V_1 \langle,\rangle_1)$ and $(V_2 \langle,\rangle_2)$ be two completions of this space. We regard V to be a dense linear subspace of the Hilbert space $(V_1, \|\cdot\|_1)$ and a dense linear subspace of the Hilbert space $(V_2, \|\cdot\|_2)$. As above we define a one-to-one, linear map $T : V_1 \to V_2$ that is onto. Let $\vec{u}, \vec{v}$ be any two elements of V_1 and find $\{\vec{u}_n\}, \{\vec{v}_n\}$ in V such that $\lim \vec{u}_n = \vec{u}$ and $\lim \vec{v}_n = \vec{v}$ for $\|\cdot\|_1$. Then

$$\langle T(\vec{u}), T(\vec{v})\rangle_2 = \lim \langle T(\vec{u}_n), T(\vec{v}_n)\rangle_2$$
$$= \lim \langle \vec{u}_n, \vec{v}_n\rangle_2$$
$$= \lim \langle \vec{u}_n, \vec{v}_n\rangle_1 = \langle \vec{u}, \vec{v}\rangle_1$$

showing that T preserves the inner product.

Now that we have proved Theorem 3 we shall speak of *the* completion of a normed or inner product space. We shall also identify the space with its "copy" in its completion. Thus we shall say that any normed space is a dense linear subspace of its completion.

Appendix C

Solutions to the Starred Problems

Note: Throughout this appendix we shall write "iff" in place of the phrase "if, and only if."

Chapter Zero

0.1.2 We have defined (x, y) to be $\{\{x\}, \{x, y\}\}$.

  (a) $(x, y) = (y, x)$ if $x = y$
      If $x = y$, then we have $(x, y) = (x, x)$ and $(y, x) = (x, x)$, and these are the same. Now suppose $(x, y) = (y, x)$, then $\{\{x\}, \{x, y\}\} = \{\{y\}, \{y, x\}\}$. We want to show that $x = y$. Suppose this is false. Then we have $\{x\} \in \{\{y\}, \{y, x\}\}$ and $\{x\} \neq \{y, x\}$, hence $\{x\} = \{y\}$. But this gives us $x = y$ and we are done.

  (b) $(x, y) = (u, v)$ if $x = u$ and $y = v$. Assume $(x, y) = (u, v)$. Suppose first that $u = v$. Then $(u, v) = \{\{u\}, \{u, v\}\} = \{\{u\}, \{u, u\}\} = \{\{u\}, \{u\}\} = \{\{u\}\}$. Now $\{x\} \in (u, v)$ and so $\{x\} = \{u\}$, giving us $x = u$. Furthermore, (x, y) must be a singleton (because, as we have just seen (u, v) is a singleton). Hence $\{x, y\} = \{x\}$ and because $y \in \{x, y\}$, $y = x$. Thus $x = y = u = v$ in this case. Similarly, if $x = y$, then $x = y = u = v$.

  Suppose now that $x \neq y$ and hence $u \neq v$. Then both (x, y) and (u, v) contain one singleton: $\{x\}$ and $\{u\}$. Clearly then, $\{x\} = \{u\}$, giving us $u = x$. Furthermore, each ordered pair contains exactly one unordered pair: $\{x, y\}$ and $\{u, v\}$, hence $\{x, y\} = \{u, v\}$. Now $y \in \{x, y\}$ so $y \in \{u, v\}$. If $y = u$, then $y = x$, which is false. Thus $y = v$.

0.1.3 $X - Y = X \cap \overline{Y}$

Let $x\epsilon X - Y$. Then $x\epsilon X$ and $x \notin Y$. Thus $x\epsilon X$ and $x \notin X \cap Y$, hence $x\epsilon X - X \cap Y$. Thus $X - Y \subseteq X - X \cap Y$. Now let $x\epsilon X - X \cap Y$. Then $x\epsilon X$ and $x \notin X \cap Y$. Thus x cannot be in Y. So $x\epsilon X, x \notin Y$ giving us $x\epsilon X - Y$. Thus $X - X \cap Y \subseteq X - Y$.

0.1.4 (a) Show (i) $C_E(X \cup Y) = C_E(X) \cap C_E(Y)$ and (ii) $C_E(X \cap Y) = C_E(X) \cup C_E(Y)$.

  i. $x\epsilon C_E(X \cup Y)$ iff $x \notin X \cup Y$ iff $x \notin X$ and $x \notin Y$ iff $x\epsilon C_E(X)$ and $x\epsilon C_E(Y)$ iff $x\epsilon C_E(X) \cap C_E(Y)$.

  ii. $x\epsilon C_E(X \cap Y)$ iff $x \notin X \cap Y$ iff $x \notin X$ or $x \notin Y$ iff $x\epsilon C_E(X)$ or $x\epsilon C_E(Y)$ iff $x\epsilon C_E(X) \cup C_E(Y)$.

(b) $X - Y = X \cap [C_E(X \cap Y)] = X \cap C_E(Y)$

$x\epsilon X - Y$ iff $x\epsilon X$ and $x \notin Y$ iff $x\epsilon X$ and $x\epsilon C_E(Y)$ iff $x\epsilon X \cap C_E(Y)$. Thus $X - Y = X \cap C_E(Y)$. By problem 1.3, $X - Y = X - X \cap Y$ and by what we have just shown $X - X \cap Y = X \cap C_E(X \cap Y)$. Thus $X - Y = X \cap C_E(X \cap Y)$.

(c) $X \cup Y = E$ iff $\{x\epsilon E \mid x \notin X\} \subseteq Y$ iff $C_E(X) \subseteq Y$. By symmetry $X \cup Y = E$ iff $C_E(Y) \subseteq X$.

(d) If $X \subseteq Y$ and $x\epsilon C_E(Y)$, then $x \notin Y$ so $x \notin X$; i.e., if $x\epsilon C_E(Y)$, then $x\epsilon C_E(X)$, giving us $C_E(Y) \subseteq C_E(X)$. Now suppose $C_E(X) \supseteq C_E(Y)$. By what has just been shown, $C_E[C_E(X)] \subseteq C_E[C_E(Y)]$; i.e., $X \subseteq Y$.

If $X \cup Y = Y$, then for any $x\epsilon X$, $x\epsilon X \cup Y$ and so $x\epsilon Y$. Thus $X \subseteq Y$. Conversely, if $X \subseteq Y$, then any $x\epsilon X \cup Y$ must be in $X \subseteq Y$ or in Y; i.e., $X \cup Y \subseteq Y$. But clearly $Y \subseteq X \cup Y$ and so $X \cup Y = Y$.

If $X \cap Y = X$, then $x\epsilon X$ implies $x\epsilon X \cap Y$ and hence $x\epsilon Y$. Thus $X \subseteq Y$. Conversely, if $X \subseteq Y$, then for any $x\epsilon X$, $x\epsilon X$ and $x\epsilon Y$ hence $x\epsilon X \cap Y$. Thus $X \subseteq X \cap Y$. But $X \supseteq X \cap Y$ and so $X = X \cap Y$.

0.2.1 (a) Because $\equiv$ is an equivalence relation it is reflexive (property (a) of the definition). Thus, for every $x\epsilon X$, $x \equiv x$ and so $x\epsilon [x]$ showing that this set is nonempty.

(b) Suppose $z\epsilon [x] \cap [y]$. Then $x \equiv z$ and $y \equiv z$. Because an equivalence relation is symmetric, we have $x \equiv z$ and $z \equiv y$ (property (b) of the definition). But then by property (c) of the definition we must have $x \equiv y$. It follows then that $[x] = [y]$.

Now suppose $[x] \neq [y]$. Then $x \not\equiv y$ and because $z\epsilon [x] \cap [y]$ implies, as we have just shown, that $x \equiv y$, we must have $[x] \cap [y] = \phi$.

0.2.2 $f : X \to Y$, g is a partial function from Y into Z.

(a) Suppose $R(f) \subseteq D(g)$. Then for any $x \epsilon X$, $f(x) \epsilon R(f) \subseteq D(g)$ and so $g[f(x)]$ has meaning; i.e., it is a unique element of Z. Thus when $R(f) \subseteq D(g)$, $g[f(x)]$ is a well-defined function from X into Z.

(b) Let $f : X \to Y$ and suppose that there is a function $g : Y \to X$ such that $(fog)(y) = y$ for all $y \epsilon Y$ and $(gof)(x) = x$ for all $x \epsilon X$. Then f is onto because it maps the element $g(y) \epsilon X$ onto the element $y \epsilon Y$ showing that $R(f) = Y$. It is also one-to-one. For suppose $f(x) = f(x')$. Then $x = g[f(x)] = g[f(x')] = x\prime$.

Now suppose $f : X \to Y$ is given to be one-to-one and onto. Then for each $y \epsilon Y$ the set $pre_f(y)$ is a singleton. We set $g(y)$ equal to the unique element of this set. Then clearly $g[f(x)] = x$ for every $x \epsilon X$. We also have $f[g(y)] = y$, by the way g was defined, for every $y \epsilon Y$.

0.2.3 We have a set X and its power set $P(X)$.

(a) We define, for Y, Z in $P(X)$, $Y \equiv Z$ to mean there is a one-to-one mapping f from Y onto Z. Clearly, $Y \equiv Y$ for every $Y \epsilon P(X)$ for we may define $f(y) = y$ for all $y \epsilon Y$ and in this way obtain a map $f : Y \to Y$ that is one-to-one and onto. Suppose that $Y \equiv Z$. Then we have $g : Y \to Z$, g is one-to-one and onto. By problem 2(b) we know that the map $g^{-1} : Z \to Y$ exists and that it is one-to-one and onto. Thus $Z \equiv Y$ showing that our relation $\equiv$ is symmetric.

Finally, suppose $Y \equiv Z$ and $Z \equiv W$. Let $f : Y \to Z$ and $g : Z \to W$ be one-to-one, onto functions. Consider $gof : Y \to W$. This map is onto because given any $w \epsilon W$ there is a $z \epsilon Z$ such that $g(z) = w$, and there is a $y \epsilon Y$ such that $f(y) = z$, hence $(gof)(y) = w$. It is also a one-to-one mapping because $(gof)(y) = gof(y')$ implies $f(y) = f(y')$ because g is one-to-one, and this implies $y = y'$ because f is one-to-one. Thus $Y \equiv W$ showing that $\equiv$ is transitive.

(b) For Y, Z in $P(X)$ define $Y \leq Z$ to mean $Y \subseteq Z$. It is clear that for any $Y \epsilon P(X)$, $Y \subseteq Y$ and so $Y \leq Y$. Thus our relation is reflexive. Suppose that $Y \leq Z$ and $Z \leq W$. Then $Y \subseteq Z$ and $Z \subseteq W$ hence $Y \subseteq W$, which means $Y \leq W$. Thus our relation is transitive. Finally, if $Y \leq Z$ and $Z \leq Y$, then we have both $Y \subseteq Z$ and $Z \subseteq Y$, which means $Y = Z$. So our relation is antisymmetric.

0.2.5 We have a family of sets $\{A_\lambda \mid \lambda \epsilon I\}$ with $A_\lambda \subseteq X$ for each $\lambda \epsilon I$. Then

(a) $C_X \left[\bigcup_{\lambda \epsilon I} A_\lambda \right] = \bigcap_{\lambda \epsilon I} C_X(A_\lambda)$

We have $x \epsilon C_X \left[\bigcup_{\lambda \epsilon I} A_\lambda \right]$ iff $x \notin \bigcup A_\lambda$ iff $x \notin A_\lambda$ for every $\lambda \epsilon I$ iff $x \epsilon C_X(A_\lambda)$ for every $\lambda \epsilon I$ iff $x \epsilon \bigcap_{\lambda \epsilon I} C_X(A_\lambda)$.

(b) $C_X \left[\bigcap_{\lambda \epsilon I} A_\lambda \right] = \bigcup_{\lambda \epsilon I} C_X (A_\lambda)$

We have $x \epsilon C_X [\cap A_\lambda]$ if $x \notin \bigcap_{\lambda \epsilon I} A_\lambda$ iff $x \notin A_\lambda$ for some $\lambda \epsilon I$ iff $x \epsilon C_X (A_\lambda)$ for some $\lambda \epsilon I$ iff $x \epsilon \bigcup_{\lambda \epsilon I} C_X (A_\lambda)$.

0.4.1 (a) Given $z = a + bi$, $\bar{z} = a - bi$ and so $z \cdot \bar{z} = a^2 - abi + abi - b^2 i^2 = a^2 + b^2 = |z|^2$.

(b) If $z = a + bi \neq 0$, then $|z| \neq 0$. Now $z \cdot \left(\frac{\bar{z}}{|z|^2} \right) = \frac{|z|^2}{|z|^2} = 1$, thus any $z \neq 0$ has a multiplicative inverse.

Suppose we have $z \cdot w = 1$ and $z \cdot z' = 1$. Then $z' = z' \cdot 1 = z' \cdot (z \cdot w) = (z' \cdot z) \cdot w = 1 \cdot w = w$. Thus the inverse of z is unique.

0.4.3 We are given $z = r(\cos \theta + i \sin \theta)$ and $w = s(\cos \varphi + i \sin \varphi)$.

(a) $z \cdot w = r(\cos \theta + i \sin \theta) \cdot s(\cos \varphi + i \sin \varphi)$

$= rs[(\cos \theta \cos \varphi - \sin \theta \sin \varphi) + i(\sin \theta \cos \varphi + \cos \theta \sin \varphi)]$

$= rs[\cos(\theta + \varphi) + i \sin(\theta + \varphi)]$.

(b) We want to show that $z^n = r^n(\cos \theta + i \sin \theta)$ for every natural number n. Clearly, this is true when $n = 1$. Suppose that for some n, it is false. Then $\{n \notin \mathbb{N} \mid z^n \neq r^n(\cos n\theta + i \sin n\theta)\}$ is not empty. Let n_0 be the smallest member of this set. As we have already noted, $n_0 > 1$. Now $n_0 - 1$ is a natural number and, from the way n_0 was defined, $z^{n_0-1} = r^{n_0-1}[\cos(n_0 - 1)\theta + i \sin(n_0 - 1)\theta]$. But then $z^{n_0} = z \cdot z^{n_0-1} = r(\cos \theta + i \sin \theta) \cdot r^{n_0-1}[\cos(n_0 - 1)\theta + i \sin(n_0 - 1)\theta]$, giving us, by (a), $z^{n_0} = r^{n_0}(\cos n_0 \theta + i \sin n_0 \theta)$. This is a contradiction.

0.4.4 We are given two complex numbers, z, w.

(a) $z = re^{i\theta}$ and $w = se^{i\theta}$. Now $\left| e^{i\alpha} \right| = |\cos \alpha + i \sin \alpha| = (\cos^2 \alpha + \sin^2 \alpha)^{\frac{1}{2}} = 1$ for any real α. Thus $|z \cdot w| = rs = |z| \cdot |w|$.

(b) We want to show that $|z + w| \leq |z| + |w|$.

Note that $|z + w|^2 = (z + w) \cdot \overline{(z + w)} = (z + w) \cdot (\bar{z} + \bar{w})$ and hence $|z + w|^2 = |z|^2 + z\bar{w} + \bar{z} \cdot w + |w|^2$.

Now write $z = re^{i\theta}, w = se^{i\varphi}$, then $z \cdot \bar{w} + \bar{z} \cdot w = rs \left[e^{i(\theta - \varphi)} + e^{i(\varphi - \theta)} \right] = rs[\cos(\theta - \varphi) + i \sin(\theta - \varphi) + \cos(\varphi - \theta) + i \sin(\varphi - \theta)] = 2rs \cos(\theta - \varphi)$ because $\cos(\theta - \varphi) = \cos(\varphi - \theta)$ and $\sin(\theta - \varphi) = -\sin(\varphi - \theta)$. Thus $z \cdot \bar{w} + \bar{z} \cdot w$ is the real number $2rs \cos(\theta - \varphi)$, which is clearly less than or equal to $2rs = 2|z||w|$. Using this we have $|z + w|^2 = |z|^2 + z \cdot \bar{w} + \bar{z} \cdot w + |w|^2 \leq |z|^2 + 2|z||w| + |w|^2 = (|z| + |w|)^2$ and this proves our inequality.

0.4.5 Let $p(z)$ be a polynomial with complex coefficients having degree $n > 0$. For any complex number r we may divide $p(z)$ by $(z-r)$ to get $p(z) = (z-r)q(z) + \rho(z)$. Here $\rho(z) \equiv 0$ or $\deg \rho(z) < \deg(z-r)$; i.e., $\rho(z)$ must be a constant. Let r_1 be a root of the equation $p(z) = 0$. Then $\rho(r_1) = 0$, so $\rho(z) \equiv 0$ and we have $p(z) = (z-r_1)q_1(z)$. Clearly, $q_1(z)$ has degree $n-1$. If $n-1 \neq 0$, then the equation $q_1(z) = 0$ has a complex root r_2. As before $q_1(z) = (z-r_2)q_2(z)$ where $q_2(z)$ has degree $n-2$. Thus $p(z) = (z-r_1)(z-r_2)q_2(z)$. Continue in this way. The process ends when we have $p(z) = (z-r_1)\cdots(z-r_n)q_n(z)$ because then the degree of $q_n(z)$ must be zero; i.e., $q_n(z)$ must be nonzero constant.

0.4.6 For any $z = a+bi$ in $\mathbb{C}$, $u(z) = a$ and $v(z) = b$. Now let $f : \mathbb{R} \to \mathbb{C}$ be given. Then $f(x) = a+bi = u(z) + v(z)i = (uof)(x) + (vof)(x)i$, and we may set $g(x) = (uof)(x)$, $h(x) = (vof)(x)$ for all $x\epsilon\mathbb{R}$ obtaining two functions from $\mathbb{R}$ to $\mathbb{R}$ such that $f(x) = g(x) + ih(x)$ for all x.

0.5.2 We are told that f is differentiable at $s_0\epsilon\mathbb{K}$. We want to show that f is continuous at this point; i.e., we want to show that $\lim_{s \to s_0} f(s) = f(s_0)$. The trick is to write

$$f(s) = \left[\frac{f(s) - f(s_0)}{s - s_0}\right](s - s_0) + f(s_0)$$

Taking the limit as s approaches s_0 the quantity in the square brackets tends to $f'(s_0)$ and hence $\lim_{s-s_0} f(s) = f(s_0)$ as was to be shown.

0.5.3 We are given $f : \mathbb{R} \to \mathbb{C}$ and we write $f(x) = g(x) + ih(x)$ where g, h map to $\mathbb{R}$ to $\mathbb{R}_+$ (problem 0.4.6).

(a) $|f(x) - (u_0 + iv_0)|^2 = |(g(x) - u_0) + i(h(x) - v_0)|^2 = |g(x) - u_0|^2 + |h(x) - v_0|^2$. Now if $\lim_{x-x_0} f(x) = u_0 + iv_0$, then $\lim_{x-x_0} g(x) = u_0$ and $\lim_{x \to x_0} h(x) = v_0$. Conversely, if $\lim g(x) = u_0$ and $\lim h(x) = v_0$, then $\lim f(x) = u_0 + iv_0$.

(b) f is continuous on $[a,b]$ iff both g and h are continuous on this interval. This follows immediately from (a) and the definitions.

(c)

$$\left|\frac{f(x) - f(x_0)}{x - x_0} - (u_0 + iv_0)\right|^2 = \left|\frac{g(x) - g(x_0)}{x - x_0} - u_0\right|^2 + \left|\frac{h(x) - h(x_0)}{x - x_0} - v_0\right|^2$$

Thus if $\lim_{x \to x_0} \frac{f(x) - f(x_0)}{x - x_0} = u_0 + iv_0$, then g and h are differentiable and $g'(x_0) = u_0$, $h'(x_0) = v_0$. Conversely, if $g'(x_0) = u_0$, $h'(x_0) = v_0$, then $f'(x_0)$ exists and is equal to $u_0 + iv_0$.

0.5.7 (a) Suppose $\lim_{n \to \infty} s_n = s_0$ and let $\{s_{n_k}\}$ be any subsequence of $\{s_n\}$. Then given $\varepsilon > 0$ there is an N such that $|s_0 - s_n| < \varepsilon$ whenever $n \geq N$. Also, there is a k_0 such that $n_k \geq N$ when $k \geq k_0$. It follows that $|s_0 - s_{n_k}| < \varepsilon$ whenever $k \geq k_0$, showing that $\lim_{k \to \infty} s_{n_k} = s_0$.

(b) Suppose that $\{s_n\}$ is a Cauchy sequence. Then given $\varepsilon > 0$ there is an N that is $|s_n - s_m| < \varepsilon$ whenever both m and n exceed N. If $\{s_{n_k}\}$ is any subsequence of $\{s_n\}$, then there is a k_0 such that $n_k \geq N$ when $k \geq k_0$. It follows that $|s_{n_k} - s_{n_l}| < \varepsilon$ whenever k and l exceed k_0.

(c) Let $\{s_n\}$ be a Cauchy sequence and let $\{s_{n_k}\}$ be a subsequence such that $\lim_{k \to \infty} s_{n_k} = s_0$. Then given $\varepsilon > 0$ there is a k_0 such that $|s_{n_k} - s_0| < \frac{\varepsilon}{2}$ when $k \geq k_0$. We can also find an N such that $|s_n - s_m| < \frac{\varepsilon}{2}$ when m and n both exceed N. Now if $n \geq N$ we can choose $k \geq k_0$ so that $n_k > N$. Then

$$|s_0 - s_n| \leq |s_0 - s_{n_k}| + |s_{n_k} - s_n| < \frac{\varepsilon}{2} + \frac{\varepsilon}{2} = \varepsilon$$

Hence $\lim_{n \to \infty} s_n = s_0$.

(d) Let $\{s_n\}$ be a Cauchy sequence. Then we can choose N so that $|s_n - s_m| < 1$ when $m, n > N$. In particular, we have $|s_n| \leq |s_n - s_N| + |s_N| \leq 1 + |s_N|$ for all $n \geq N$. Let K be an integer that is greater than $|s_1|, |s_2|, ..., |s_N|$. Then we have $|s_n| \leq 1 + K$ for all n.

Chapter One

1.1.1 We have $u = u(x, y)$ and $x = r \cos \theta$, $y = r \sin \theta$.

$$\frac{\partial u}{\partial r} = \frac{\partial u}{\partial x} \frac{\partial x}{\partial r} + \frac{\partial u}{\partial y} \frac{\partial y}{\partial r} = \frac{\partial u}{\partial x} \cos \theta + \frac{\partial u}{\partial y} \sin \theta$$

$$\frac{\partial^2 u}{\partial r^2} = \cos \theta \left[\frac{\partial^2 u}{\partial x^2} \frac{\partial x}{\partial r} + \frac{\partial^2 u}{\partial y \partial x} \frac{\partial y}{\partial r} \right] + \sin \theta \left[\frac{\partial^2 u}{\partial x \partial y} \frac{\partial x}{\partial r} + \frac{\partial^2 u}{\partial y^2} \frac{\partial y}{\partial r} \right]$$

$$= \frac{\partial^2 u}{\partial x^2} \cos^2 \theta + 2 \sin \theta \cos \theta \frac{\partial^2 u}{\partial x \partial y} + \frac{\partial^2 u}{\partial y^2} \sin^2 \theta$$

because

$$\frac{\partial^2 u}{\partial x \partial y} = \frac{\partial^2 u}{\partial y \partial x}$$

$$\frac{\partial u}{\partial \theta} = \frac{\partial u}{\partial x}\frac{\partial x}{\partial \theta} + \frac{\partial u}{\partial y}\frac{\partial y}{\partial \theta} = \frac{\partial u}{\partial x}(-r\sin\theta) + \frac{\partial u}{\partial y}(r\cos\theta)$$

$$\frac{\partial^2 u}{\partial \theta^2} = \frac{\partial u}{\partial x}(-r\cos\theta) + \frac{\partial u}{\partial y}(-r\sin\theta) + (-r\sin\theta)\left[\frac{\partial^2 u}{\partial x^2}\frac{\partial x}{\partial \theta} + \frac{\partial^2 u}{\partial y \partial x}\frac{\partial y}{\partial \theta}\right]$$
$$+ (r\cos\theta)\left[\frac{\partial^2 u}{\partial x \partial y}\frac{\partial x}{\partial \theta} + \frac{\partial^2 u}{\partial y^2}\frac{\partial y}{\partial \theta}\right]$$
$$= -r\cos\theta\frac{\partial u}{\partial x} - r\sin\theta\frac{\partial u}{\partial y} + r^2\sin^2\theta\frac{\partial^2 u}{\partial x^2} - 2r^2\sin\theta\cos\theta\frac{\partial^2 u}{\partial x \partial y}$$
$$+ r^2\cos^2\theta\frac{\partial^2 u}{\partial y^2}$$

now

$$\frac{\partial^2 u}{\partial r^2} + \frac{1}{r}\frac{\partial u}{\partial r} + \frac{1}{r^2}\frac{\partial^2 u}{\partial \theta^2} = \frac{\partial^2 u}{\partial x^2} + \frac{\partial^2 u}{\partial y^2}$$

1.2.3 $f : \mathbb{R} \to \mathbb{R}$ is continuous and $\alpha > 0$ is any element of $P(f)$. We distinguish several cases:

(a) $\alpha > 0$

Case 1. $0 < x$. Then

$$\int_x^{x+\alpha} = \int_0^{x+\alpha} - \int_0^x = \int_0^\alpha + \int_\alpha^{x+\alpha} - \int_0^x$$

Now

$$\int_\alpha^{x+\alpha} f(t)\,dt = \int_0^x f(u+\alpha)\,du = \int_0^x f(u)\,du$$

where we set $t = u + \alpha$ and used the fact that $f(u+\alpha) = f(u)$. It now follows that

$$\int_x^{x+\alpha} = \int_0^\alpha$$

Case 2. $x < 0$. Here we have two subcases:

i.   —————|————————|————|————
     $-x$ 0 $\alpha-x$ α

$$\int_{-x}^{-x+\alpha} = \int_{-x}^{0} + \int_{0}^{\alpha-x} = \int_{-x}^{0} + \int_{0}^{\alpha} - \int_{\alpha-x}^{\alpha}$$

But

$$\int_{\alpha-x}^{\alpha} f(t)\,dt = \int_{-x}^{0} f(u+\alpha)\,du = \int_{-x}^{0} f(u)\,du$$

hence $\int_{-x}^{-x+a} = \int_{0}^{\alpha}$

ii. —————|————|————————|————
     $-x$ $-x+\alpha$ 0 α

$$\int_{-x}^{-x+\alpha} = \int_{-x}^{0} - \int_{-x+\alpha}^{0} = \int_{-x}^{0} - \left[\int_{-x+\alpha}^{\alpha} - \int_{0}^{\alpha}\right]$$

Again

$$\int_{-x+\alpha}^{\alpha} = \int_{-x}^{0} \quad \text{so} \quad \int_{-x}^{-x+\alpha} = \int_{0}^{\alpha}$$

(b) $\alpha < 0$
Then

$$\int_{x-\alpha}^{x} = \int_{y}^{y+\alpha} = \int_{0}^{a}$$

where we have set $y = x - \alpha$.

1.2.4 We have $g(x) = a\cos\lambda x + b\sin\lambda x$, where a, b are complex numbers not both zero and λ is a positive constant.

It is clear that $P(g) \supseteq \{n\left(\frac{2\pi}{\lambda}\right) \mid n \in Z\}$, and if $a = 0$ (or if $b = 0$), these two sets are equal. We want to show that these sets are equal when neither a nor b is zero.

Let $\alpha \in P(g)$, so that $g(x+\alpha) = g(x)$ for all x. Then $a\cos\lambda(x+\alpha) + b\sin\lambda(x+\alpha) = a\cos\lambda x + b\sin\lambda x$ for all x. A little algebra gives us:

$$[a\cos\lambda\alpha + b\sin\lambda\alpha - a]\cos\lambda x + [b\cos\lambda\alpha - a\sin\lambda\alpha - b]\sin\lambda x = 0$$

for all x.

We need a simple result:

Lemma. If $A\cos\lambda x + B\sin\lambda x = 0$ for all x, then $A = B = 0$.

Proof. Set $x = 0$ to see that $A = 0$. Then $B\sin\lambda x = 0$ for all x. Clearly, $B = 0$ (take, for instance, $x = \frac{\pi}{2\lambda}$). This proves the lemma.

We can now conclude that $a(\cos\lambda\alpha - 1) + b\sin\lambda\alpha = 0$ and that $b(\cos\lambda\alpha - 1) - a\sin\lambda\alpha = 0$. Multiply the first of these by a, the second by b, and add the results to obtain $(a^2 + b^2)(\cos\lambda\alpha - 1) = 0$.

If $a^2 + b^2 \neq 0$, then $\cos\lambda\alpha = 1$; hence $\lambda\alpha = 2k\pi$ and $\alpha = \frac{2k\pi}{\lambda}$, where $k \in \mathbb{Z}$.

If $a^2 + b^2 = 0$, then $a = \pm ib$ and clearly $b \neq 0$ (otherwise we'd have both a and b zero). Thus when $a = ib$ our equation becomes $ib(\cos\lambda\alpha - 1) + b\sin\lambda\alpha = 0$ or $i(\cos\lambda\alpha - 1) + \sin\lambda\alpha = 0$. Equating real and imaginary parts gives $\cos\lambda\alpha = 1$. The case $a = -bi$ is similar.

1.2.6 Given $f : \mathbb{R} \to \mathbb{C}$ there are functions g, h that map $\mathbb{R}$ to $\mathbb{R}$ such that $f(x) = g(x) + ih(x)$ for all x (see Chapter Zero, Exercises 4, problem 6). We must show that (a) $P(f) = P(g) \cap P(h)$.

It is clear that $P(g) \cap P(h) \subseteq P(f)$. Suppose $\alpha \in P(f)$. Then $f(x + \alpha) = f(x)$ for all x, giving $g(x + \alpha) + ih(x + \alpha) = g(x) + ih(x)$ for all x. Equating real and imaginary parts we see that $\alpha \in P(g)$ and $\alpha \in P(h)$.

(b) If f is periodic, then $P(f)$ contains a nonzero member. Thus $P(g)$ and $P(h)$ both contain this nonzero number showing that g and h are periodic. To see that the converse is false, consider $f(x) = \sin 2\pi x + i\sin(2\pi\sqrt{2})x$. Then $P(g) = \{n \mid n \in \mathbb{Z}\}$ and $P(h) = \left\{\frac{n}{\sqrt{2}} \mid n \in \mathbb{Z}\right\}$. Clearly, $P(g) \cap P(h) = \{0\}$; hence $P(f) = \{0\}$ and f is not periodic.

(c) If f is continuous, periodic, and nonconstant, then f has a smallest positive period.

We have seen that f is continuous iff g and h are (see Chapter Zero, Exercises 5, problem 3(b)). If f does not have a smallest positive period, then $P(f)$ contains arbitrarily small positive numbers. But because $P(f) = P(g) \cap P(h) \subseteq P(g)$ we see that $P(g)$, and similarly $P(h)$, contains arbitrarily small positive numbers. It follows from Lemma 2 of Chapter One, Section 2 that both g and h are constant, and this is a contradiction.

1.2.7 Let $f : \mathbb{R} \to \mathbb{C}$ be continuous and periodic. Then f is uniformly continuous on $\mathbb{R}$.

Let $\varepsilon > 0$ be given. We must show that we can find a $\delta > 0$ such that $|x - y| < \delta$ implies $|f(x) - f(y)| < \varepsilon$. Let $\alpha \in P(f)$, $\alpha > 0$.

Then by Theorem 3 of Chapter Zero, Section 5 we know that f is uniformly continuous on $[0, \alpha]$. Thus, for the given ε, we can find δ so that $|f(x) - f(y)| < \varepsilon$ whenever x, y are in $[0, \alpha]$ and $|x - y| < \delta$. Clearly, we may suppose that $\delta < \alpha$.

Now let $x, y \in \mathbb{R}$ be such that $|x - y| < \delta$. We may suppose $x < y$. Consider $\{[n\alpha, (n+1)\alpha] \mid n \in \mathbb{Z}\}$. There are two cases:

(i) $n\alpha \leq x < y \leq (n+1)\alpha$ for some n.
Then $|f(y) - f(x)| = |f(y - n\alpha) - f(x - n\alpha)| < \varepsilon$ because $x - n\alpha$ and $y - n\alpha$ are in $[0, \alpha]$ and $|(x - n\alpha) - (y - n\alpha)| < \delta$.

(ii) $x < n\alpha < y < (n+1)\alpha$ for some n.
Then $|f(y) - f(x)| \leq |f(y) - f(n\alpha)| + |f(n\alpha) - f(x)| < 2\varepsilon$ because $n\alpha$ and y are in $[n\alpha, (n+1)\alpha]$ and $|y - n\alpha| < \delta$, and $n\alpha$ and x are in $[(n-1)\alpha, n\alpha]$ and $|n\alpha - x| < \delta$.

1.3.1 Here λ is a fixed positive number.

(a) If $ae^{\lambda x} + be^{-\lambda x} = 0$ for all x, then $a = b = 0$.
Setting $x = 0$ gives the equation $a + b = 0$. Differentiating and setting $x = 0$ in the result gives $\lambda(a - b) = 0$. Solving these equations we see that $a = b = 0$.

(b) Set $f(x) = c_1 e^{\lambda x} + c_2 e^{-\lambda x}$ where c_1 and c_2 are constants not both zero. Suppose $f(x + \alpha) = f(x)$ for all x. Then $0 = c_1 \left(e^{\alpha\lambda} - 1\right) e^{\lambda x} + c_2 \left(e^{-\alpha\lambda} - 1\right) e^{-\lambda x}$ for all x. It follows from part (a) that $c_1 \left(e^{\alpha\lambda} - 1\right) = 0 = c_2 (e^{-\alpha x} - 1)$. If $c_1 \neq 0$, then $e^{\alpha\lambda} = 1$, giving us $\alpha\lambda = 0$. Because $\lambda > 0$, $\alpha = 0$. If $c_2 \neq 0$, then $e^{-\alpha\lambda} = 1$, and again we get $\alpha = 0$. Thus $P(f) = \{0\}$ and the function f is not periodic.

1.3.2 Equation (7) of Chapter One, Section 3 is $r^2 R''(r) + r R'(r) - n^2 R(r) = 0$. Set $r = e^t$. Then $\frac{dR}{dt} = \frac{dR}{dr} \cdot \frac{dr}{dt}$ or, demoting the derivative with respect to t by $\dot{R}$, $\dot{R} = R' e^t$.

Now $\ddot{R} = e^t \left[\frac{d}{dr} R' \cdot \frac{dr}{dt}\right] + R' \frac{d}{dt} e^t = e^{2t} R'' + e^t R'$. Thus $\ddot{R} - \dot{R} = e^{2t} R''$.

We now have $r^2 R'' = r^2 \left[e^{-2t} \left(\ddot{R} - \dot{R}\right)\right] = \ddot{R} - \dot{R}$ and $rR' = e^t \left(e^{-t}\dot{R}\right) = \dot{R}$. Using these equations (7) becomes $\ddot{R} - n^2 R = 0$. The general solution to this last equation is $R = c_1 e^{nt} + c_2 e^{-nt}$ and because $r = e^t$ we see that $R = c_1 r^n + c_2 r^{-n}$.

1.3.3 We want to solve $r^2 R''(r) + rR'(r) = 0$. Again we use $r = e^t$. We saw in the last problem (1.3.2) that $e^{2t} R'' = \ddot{R} - \dot{R}$ and $e^t R' = \dot{R}$. Using these, our equation becomes $\ddot{R} = 0$. Hence $\dot{R} = c$ and $R = ct + d$ where c and d are constants. Because $r = e^t$, $R(r) = c \ln r + d$.

1.4.1 We have a sequence $\{u_n(r,\theta)\}_{n=1}^{\infty}$ of functions that are continuous on $G \cup T$ and harmonic in G. If $\{u_n(1,\theta)\}$ is uniformly convergent over T, then $\{u_n(r,\theta)\}$ is uniformly convergent over $G \cup T$.

Given $\varepsilon > 0$ we must show that we can find an integer N such that $|u_n(r,\theta) - u_m(r,\theta)| < \varepsilon$ for all $(r,\theta) \in G \cup T$, whenever $m, n \geq N$ (Lemma 1 of Chapter One, Section 4). Now from our hypothesis we can find, given $\varepsilon > 0$, N so that $|u_n(1,\theta) - u_m(1,\theta)| < \varepsilon$ for all $\theta \in T$ whenever $m, n \geq N$. Thus $-\varepsilon < u_n(1,\theta) - u_m(1,\theta) < \varepsilon$ for all $\theta \in T$ because our functions are real-valued. Because they are also continuous on T, we have

$$-\varepsilon < \min_{\theta \in T}\{u_n(1,\theta) - u_m(1,\theta)\} \leq \max_{\theta \in T}\{u_n(1,\theta) - u_m(1,\theta)\} < \varepsilon$$

whenever $m, n \geq N$. But by Theorem 1 of Chapter One, Section 1 we have

$$-\varepsilon < \min_{\theta \in T}\{u_n(1,\theta) - u_m(1,\theta)\} \leq \min_{(r,\theta) \in G \cup T}\{u_n(r,\theta) - u_m(r,\theta)\}$$
$$\leq \max_{(r,\theta) \in G \cup T}\{u_n(r,\theta) - u_m(r,\theta)\}$$
$$\leq \max_{\theta \in T}\{u_n(1,\theta) - u_m(1,\theta)\} < \varepsilon$$

whenever $m, n \geq N$.

1.4.2 (a) For $n \in \mathbb{N}$, n fixed, $u_n(r,\theta) = r^n(a_n \cos n\theta + b_n \sin n\theta)$ is harmonic in G, i.e., inside the unit circle.

$$\frac{\partial u_n}{\partial r} = nr^{n-1}(a_n \cos\theta + b_n \sin n\theta),$$

$$\frac{\partial^2 u_n}{\partial r^2} = n(n-1)r^{n-2}(a_n \cos n\theta + b_n \sin n\theta)$$

$$\frac{\partial u_n}{\partial \theta} = r^n(-na_n \sin n\theta + nb_n \cos n\theta), \frac{\partial^2 u_n}{\partial \theta^2}$$
$$= r^n(-n^2 a_n \cos n\theta - n^2 b_n \sin n\theta)$$

$$\frac{\partial^2 u_n}{\partial r^2} + \frac{1}{r}\frac{\partial u_n}{\partial r} + \frac{1}{r^2}\frac{\partial^2 u_n}{\partial \theta^2} = n(n-1)r^{n-1}(a_n \cos n\theta$$
$$+ b_n \sin n\theta) + nr^{n-2}(a_n \cos n\theta + b_n \sin n\theta)$$
$$- n^2 r^{n-2}(a_n \cos n\theta - b_n \sin \theta)$$
$$= [n(n-1) + n - n^2] r^{n-2}[a_n \cos n\theta - b_n \sin \theta]$$
$$= 0 \quad \text{for all } (r,\theta) \in G$$

(b) For any fixed ρ, $0 < \rho < 1$, the series $\sum_{n=0}^{\infty} \rho^n$, $\sum_{n=1}^{\infty} n\rho^n$, $\sum_{n=1}^{\infty} n^2 \rho^n$ are convergent.

Because these are series with positive terms, we can apply the ratio test. Clearly,

$$\rho = \lim \frac{\rho^{n+1}}{\rho^n} = \lim \frac{(n+1)\rho^{n+1}}{n\rho^n} = \lim \frac{(n+1)^2 \rho^{n+1}}{n^2 \rho^n}$$

Because we are given that $\rho < 1$, all three series converge.

(c) Suppose that for some M, $|a_n| \leq M$ and $|b_n| \leq M$ for all n. Then $u(r,\theta) = \sum_{n=1}^{\infty} r^n (a_n \cos n\theta + b_n \sin n\theta)$ is harmonic in G, i.e., for $0 \leq \theta \leq 2\pi$ and $r < 1$.

$|r^n(a_n \cos n\theta + b_n \sin n\theta)| \leq r^n 2M$ for every n. Because $\sum r^n$ is convergent, the M-test (Corollary 1 to Lemma 1, Chapter One, Section 4) shows that the series given converges uniformly in G. Similarly, each of the series

$$\sum \frac{\partial}{\partial r} r^n (a_n \cos n\theta + b_n \sin n\theta) = \sum nr^{n-1}(a_n \cos n\theta + b_n \sin n\theta)$$

$$\sum \frac{\partial^2}{\partial r^2} r^n (a_n \cos n\theta + b_n \sin n\theta) = \sum n(n-1)r^{n-2}(a_n \cos n\theta + b_n \sin n\theta)$$

$$\sum \frac{\partial}{\partial \theta} r^n (a_n \cos n\theta + b_n \sin n\theta) = \sum nr^n(-a_n \sin n\theta + b_n \cos n\theta)$$

$$\sum \frac{\partial^2}{\partial \theta^2} r^n (a_n \cos n\theta + b_n \sin n\theta) = \sum n^2 r^n(-a_n \cos n\theta - b_n \sin n\theta)$$

is uniformly convergent. Writing $u_n(r,\theta)$ for $r^n(a_n \cos n\theta + b_n \sin n\theta)$ we may use Lemma 4 of Chapter One, Section 4 to obtain

$$\frac{\partial^2 u}{\partial r^2} + \frac{1}{r}\frac{\partial u}{\partial r} + \frac{1}{r^2}\frac{\partial^2 u}{\partial \theta^2} = \sum \left(\frac{\partial^2 u_n}{\partial r^2} + \frac{1}{r}\frac{\partial u_n}{\partial r} + \frac{1}{r^2}\frac{\partial^2 u_n}{\partial \theta^2} \right)$$

which is zero by (a).

1.5.2 If $f \subset C_r(T)$, then there is an M such that $|a_n(f)| \leq M$ and $|b_n(f)| \leq M$ for every n.

Because f is continuous on T there is a number μ such that $|f(\theta)| \leq \mu$ for all $\theta \in T$. Thus because $a_n(f) = \frac{1}{\pi} \int_{-\pi}^{\pi} f(\theta) \cos n\theta \, d\theta$ we have $|a_n(f)| \leq \frac{1}{\pi} \int |f(\theta)| \, d\theta \leq \frac{\mu}{\pi}(2\pi) = 2\mu$ for $n = 0, 1, 2, \ldots$ and because $b_n(f) = \frac{1}{\pi} \int_{-\pi}^{\pi} f(\theta) \sin n\theta \, d\theta$ we have $|b_n(f)| \leq 2\mu$, $n = 1, 2, \ldots$.

1.6.2 Suppose that the sequence $\{a_n\}_{n=1}^{\infty}$ is convergent to b. We claim that the numbers $\sigma_n = \frac{a_1+a_2+...+a_n}{n}$, $n = 1, 2, ...$ also converge to b.

$$|\sigma - b| = \left|\frac{a_1 + a_2 + ... + a_n - nb}{n}\right| \leq \frac{|a_1 - b| + ... + |a_n - b|}{n}$$

Now given $\varepsilon > 0$ we may choose N so that $|a_n - b| < \frac{\varepsilon}{2}$ when $n \geq N$. Then

$$|\sigma - b| \leq \frac{|a_1 - b| + ... + |a_{N-1} - b|}{n} + \frac{|a_N - b| + ... + |a_n - b|}{n}$$
$$\leq \frac{|a_1 - b| + ... + |a_{N-1} - b|}{n} + \frac{(n - N + 1)}{n}\frac{\varepsilon}{2}$$

We note that $\frac{n-N+1}{n} \leq 1$ and that by choosing $n > N$ sufficiently large we can make the first term less than $\frac{\varepsilon}{2}$; because the numerator of this term is fixed.

1.6.3

$$K_m(x) = \frac{1}{m}\left[\sum_{0}^{0} e^{ikx} + \sum_{-1}^{1} e^{ikx} + ... + \sum_{-(m-1)}^{m-1} e^{ikx}\right]$$
$$= \frac{1}{m}\left[me^{i(0)x} + (m-1)e^{ix} + (m-1)e^{-ix} + ... + e^{i(m-1)x}\right]$$
$$= \left(1 - \frac{0}{m}\right)e^{i(0)x} + \left(1 - \frac{1}{m}\right)e^{ix} + \left(1 - \frac{1}{m}\right)e^{-ix} + ...$$
$$+ \left(1 - \frac{m-1}{m}\right)e^{i(m-1)x}$$
$$= \sum_{k=-(m-1)}^{m-1}\left(1 - \frac{|k|}{m}\right)e^{ikx} = \sum_{k=-(m-1)}^{m-1}\left(1 - \frac{|k|}{m}\right)e^{-ikx}$$
$$= \sum_{|k|<m}\left(1 - \frac{|k|}{m}\right)e^{-ikx}$$

as was to be shown.

1.7.1 For any $f \in C_r(T)$ there is a sequence of trigonometric polynomials that converges to f uniformly over T.

Given any $n \in \mathbb{N}$ we can find (Corollary 2 to Theorem 1, Chapter One, Section 7) a trigonometric polynomial τ_n such that $|f(\theta) - \tau_n(\theta)| < \frac{1}{n}$ for all $\theta \in T$. The sequence $\{\tau_n\}$ chosen in this way converges to f uniformly over T; i.e. given $\varepsilon > 0$ we choose $N \in \mathbb{N}$ such that $\frac{1}{N} < \varepsilon$ and note that $|f(\theta) - \tau_n(\theta)| < \frac{1}{n} < \varepsilon$ for all $\theta \in T$ whenever $n \geq N$.

1.7.2 (a) Let $\tau(\theta) = \sum_{j=1}^{k} c_j e^{in_j \theta}$. Then $\int_{-\pi}^{\pi} \tau(\theta) e^{-in\theta} d\theta = 0$ unless $n = n_j$ for some j. When $n = n_j$ the integral is 2π. Hence the Euler formulas (Chapter One, Section 5) show that the complex Fourier series of $\tau(\theta)$ is $\tau(\theta)$.

(b) Given $\tau(\varphi) = \sum_{j=1}^{l} c_j e^{in_j \varphi} = \sum_{j=1}^{l} c_j (\cos n_j \varphi + i \sin n_j \varphi)$ and, as in part (a), the Euler formulas show that this last sum is the real Fourier series of $\tau(\varphi)$.

1.7.3

$$\frac{1}{1-re^{i\omega}} = \frac{1}{1-r(\cos\omega + i\sin\omega)}$$

$$= \frac{1}{1-r\cos\omega - ir\sin\omega} \cdot \frac{(1-r\cos\omega)+ir\sin\omega}{(1-r\cos\omega)+ir\sin\omega}$$

$$= \frac{1-r\cos\omega + ir\sin\omega}{(1-r\cos\omega)^2 + (r\sin\omega)^2}$$

$$= \frac{1-r\cos\omega + ir\sin\omega}{1 - 2r\cos\omega + r^2(\cos^2\omega + \sin^2\omega)}$$

$$= \frac{1-r\cos\omega}{1-2r\cos\omega+r^2} + i\left(\frac{r\sin\omega}{1-2r\cos\omega+r^2}\right)$$

Now $-1 + 2\mathrm{Re}\left[\dfrac{1}{1-re^{i\omega}}\right] = -1 + \dfrac{2(1-r\cos\omega)}{1-2r\cos\omega+r^2}$

$$= \frac{-1+2r\cos\omega - r^2 + 2 - 2r\cos\omega}{1-2r\cos\omega+r^2}$$

$$= \frac{1-r^2}{1-2r\cos\omega+r^2}$$

1.7.6 If $f \in C_r(T)$ and $c_n(f) = 0$ for all n, then the partial sums, and hence the Ceràro means, of the Fourier series of f are all identically zero. Because these means converge to f uniformly over T, f must be identically zero.

Chapter Two

2.3.1

$$\lim_{x \to 0} \frac{x}{2\sin\frac{1}{2}x} = \lim_{x \to 0} \frac{\frac{x}{2}}{\sin\left(\frac{x}{2}\right)} = \lim_{x \to 0} \frac{\frac{1}{2}}{\frac{1}{2}\cos\left(\frac{x}{2}\right)} = 1$$

$$f'_+(0) = \lim_{h \to 0} \frac{f(0+h) - f(0+0)}{h}$$

where $f(0+0) = \lim_{h \to 0^+} f(0+h)$

and in this case $f(0+0) = 1$. Thus

$$f'_+(0) = \lim_{h \to 0^+} \frac{\frac{h/2}{\sin(h/2)} - 1}{h} = \lim_{h \to 0^+} \frac{\frac{h}{2} - \sin\left(\frac{h}{2}\right)}{h \sin\left(\frac{h}{2}\right)}$$

$$= \lim_{h \to 0^+} \frac{\frac{1}{2} - \frac{1}{2}\cos\left(\frac{h}{2}\right)}{\frac{h}{2}\cos\left(\frac{h}{2}\right) + \sin\left(\frac{h}{2}\right)}$$

$$= \lim_{h \to 0^+} \frac{\frac{1}{4}\sin\left(\frac{h}{2}\right)}{\frac{-h}{4}\sin\left(\frac{h}{2}\right) + \frac{1}{2}\cos\left(\frac{h}{2}\right) + \frac{1}{2}\cos\left(\frac{h}{2}\right)} = 0$$

Similarly, $f'_-(0) = 0$.

2.3.4 (b) We are given f uniformly continuous on (a, b).

  i. Let $\{x_n\}$ be a Cauchy sequence in (a, b). We claim that $\{f(x_n)\}$ is a Cauchy sequence. Let $\alpha > 0$ be given. Because f is uniformly continuous, we can choose $\delta > 0$ so that $|x - y| < \delta$ implies $|f(x) - f(y)| < \varepsilon$. Because $\{x_n\}$ is a Cauchy sequence, we can find N such that $|x_n - x_m| < \delta$ when $m, n \geq N$. But then $|f(x_n) - f(x_m)| < \alpha$ when $m, n \geq N$, and we are done.

  ii. Now let $\{x_n\} \subseteq (a, b)$ and suppose $\lim_{n \to \infty} x_n = a$. Because $\{f(x_n)\}$ is a Cauchy sequence it converges to, say α. We want to show that $\lim_{h \to 0} f(a + h) = \alpha$; i.e., given $\varepsilon > 0$ we must show that there is a $\delta > 0$ such that
  $|f(a + h) - \alpha| < \varepsilon$ when $0 < h < \delta$. First use the uniform continuity of f to find δ so that $|f(x) - f(y)| < \frac{\varepsilon}{2}$ when $|x - y| < \delta$. Now $|f(a + h) - \alpha| \leq |f(a + h) - f(x_n)| + |f(x_n) - \alpha|$. Choose N such that $a < x_N < a + h$ and $|f(x_N) - \alpha| < \frac{\varepsilon}{2}$. Note that when $0 < h < \delta$ we have $|(a + h) - x_N| < \delta$ and hence $|f(a + h) - f(x_N)| < \frac{\varepsilon}{2}$. Thus $0 < h < \delta$ gives $|f(a + h) - \alpha| < \varepsilon$ and we are done. The proof that $\lim_{h \to 0^+} f(b - h)$ exists is similar.

Chapter Three

3.1.1 For any $\vec{x}, \vec{y}$ in E, $\vec{x} = \vec{x} - \vec{y} + \vec{y}$ and so $\|\vec{x}\| \leq \|\vec{x} - \vec{y}\| + \|\vec{y}\|$ or $\|\vec{x}\| - \|\vec{y}\| \leq \|\vec{x} - \vec{y}\|$. Because $\vec{y} = \vec{y} - \vec{x} + \vec{x}$, $\|\vec{y}\| - \|\vec{x}\| \leq \|\vec{y} - \vec{x}\|$. But $\|\vec{x} - \vec{y}\| = \|(-1)(\vec{y} - \vec{x})\| = |(-1)| \|\vec{y} - \vec{x}\|$. Thus $\|\vec{x} - \vec{y}\|$ is greater than, or equal to, both $\|\vec{x}\| - \|\vec{y}\|$ and $\|\vec{y}\| - \|\vec{x}\|$, and so we have
$$\left|\|\vec{x}\| - \|\vec{y}\|\right| \leq \|\vec{x} - \vec{y}\|.$$

3.1.2 The vector $(1, i) \in C^2$ has the property that $|1|^2 + |i|^2 = 2$, whereas $1^2 + i^2 = 0$.

3.1.3 (a) Let $y \in \theta_\varepsilon(\vec{x}_0)$. Then $\|\vec{x}_0 - \vec{y}\| < \varepsilon$. Let us take δ to be less than the smaller of the numbers $\|\vec{x}_0 - \vec{y}\|$, $\varepsilon - \|\vec{x}_0 - \vec{y}\|$. Note that with this choice $\delta + \|\vec{x}_0 - \vec{y}\| < \varepsilon$. Suppose $\|\vec{z} - \vec{y}\| < \delta$. Then $\|\vec{z} - \vec{x}_0\| \le \|\vec{z} - \vec{y}\| + \|\vec{y} - \vec{x}_0\| < \delta + \|\vec{y} - \vec{x}_0\| < \varepsilon$. Thus $\theta_\delta(\vec{y}) \subseteq \theta_\varepsilon(\vec{x}_0)$.

(b) Suppose $\vec{y} \notin B_\varepsilon(\vec{x}_0)$. Then $\|\vec{y} - \vec{x}_0\| > \varepsilon$. Let us take $\delta < \|\vec{y} - \vec{x}_0\| - \varepsilon$ and consider $O_\delta(\vec{y})$. If $\vec{z} \in O_\delta(\vec{y})$, then $\|\vec{z} - \vec{x}_0\| = \|(\vec{z} - \vec{y}) - (\vec{x}_0 - \vec{y})\|$ and this, by 3.1.1, is
$$\ge \|(\vec{x}_0 - \vec{y}) - (\vec{z} - \vec{y})\| \ge \left| \|\vec{y} - \vec{x}_0\| - (\|\vec{y} - \vec{x}_0\| - \varepsilon) \right| > \varepsilon.$$
Thus $O_\delta(\vec{y}) \cap B_\varepsilon(\vec{x}_0) = \emptyset$.

(c) Let $\{O_\alpha | \alpha \in I\}$ be a collection of open sets and let $O = \bigcup_{\alpha \in I} O_\alpha$. If $\vec{x} \in O$, then $\vec{x} \in O_\alpha$ for some $\alpha \in I$. Because O_α is open there is an $\varepsilon > 0$ such that $O_\varepsilon(\vec{x}) \subseteq O_\alpha$. But then $O_\varepsilon(\vec{x}) \subseteq O$, showing that O is an open set. Now let $\{O_j\}_{j=1}^n$ be a finite family of open sets and let $O = \bigcap_{j=1}^n O_j$. If $\vec{x} \in O$, then $\vec{x} \in O_j$ for every j. Thus because O_j is an open set, we have $\varepsilon_j > 0$ such that $O_{\varepsilon_j}(\vec{x}) \subseteq O_j$, and this is true for $j = 1, 2, ..., n$. Let $\varepsilon = \min \{\varepsilon_j\}_{j=1}^n$. Then $O_\varepsilon(\vec{x}) \subseteq O_{\varepsilon_j}(\vec{x}) \subseteq O_j$ for $j = 1, 2, , n$. Hence $O_\varepsilon(\vec{x}) \subseteq O$. If $O = \emptyset$, it is an open set by definition, for it is a neighborhood of each of its points.

(d) A set is closed iff its complement is open. Let $\{C_\alpha \mid \alpha \in I\}$ be a family of closed sets and let $C = \bigcap_{\alpha \in I} C_\alpha$. Then $E - C = E - \bigcap_{\alpha \in I} C_\alpha = \bigcup_{\alpha \in I} (E - C_\alpha)$ (Chapter Zero, Exercises 2, problem 5). Because each C_α is closed, each $E - C_\alpha$ is open and so by (c), $E - C$ is open. Thus C is closed.

Now let $\{C_j\}_{j=1}^n$ be a finite family of closed sets and let $C = \bigcup_{j=1}^n C_j$. Then $E - C = E - \bigcup_{j=1}^n C_j = \bigcap_{j=1}^n (E - C_j)$ by (Chapter Zero, Exercises 1, problem 4). Because each of the sets $E - C_j$ is open, so also is $E - C$ by (c). We conclude that C is a closed set.

3.2.1 $(E, \|\cdot\|)$ and $\{\vec{x}_n\} \subseteq E$ given, and $\{\vec{x}_n\}$ converges to $\vec{x}_0$ for all $\|\cdot\|$.

(a) By 3.1.1 we have $\left| \|\vec{x}_0\| - \|\vec{x}_n\| \right| \leq \|\vec{x}_0 - \vec{x}_n\|$. This last quantity tends to zero as $n \to \infty$, and the first quantity is nonnegative. Thus $\lim_{n\to\infty} \|\vec{x}_n\| = \|\vec{x}_0\|$.

(b) Given $\varepsilon > 0$ we choose N so that $\|x_0 - x_n\| < \frac{\varepsilon}{2}$ when $n \geq N$. Then $\|\vec{x}_n - \vec{x}_m\| \leq \|(\vec{x}_n - \vec{x}_0) + (\vec{x}_0 - \vec{x}_m)\| \leq \|\vec{x}_0 - \vec{x}_n\| + \|\vec{x}_0 - \vec{x}_m\| < \varepsilon$ when $m, n \geq N$.

3.2.2 S is a nonempty set, S' the set of adherent points of S, $\bar{S} = S \cup S'$.

(a) $\vec{x}_0 \in S'$ iff there is a sequence of distinct points of S that converges to $\vec{x}_0$. If $\{\vec{s}_n\} \subseteq S$, $\vec{s}_n \neq \vec{s}_m$ when $n \neq m$ and $\lim_{n\to\infty} \vec{s}_n = \vec{x}_0$, then given $\varepsilon > 0$ there is an N such that $\vec{s}_n \in O_\varepsilon(\vec{x}_0)$ whenever $n \geq N$. Clearly, $\{\vec{s}_n \mid n \geq N\}$ is an infinite set and because $\varepsilon > 0$ is arbitrary we see that $\vec{x}_0 \in S'$.

Now suppose that $\vec{x}_0 \in S'$. We can find $\vec{s}_1 \in O_1(\vec{x}_0) \cap S$. Next we choose $\vec{s}_2 \in O_{\frac{1}{2}}(\vec{x}_0) \cap S$ and $\vec{s}_2 \neq \vec{s}_1$. Then choose $\vec{s}_3 \in O_{\frac{1}{3}}(\vec{x}_0) \cap S$ such that $\vec{s}_3 \notin \{\vec{s}_1, \vec{s}_2\}$. Continue in this way. The sequence $\{\vec{s}_n\}$ so obtained consists of distinct points of S, and clearly this sequence converges to $\vec{x}_0$.

(b) We shall solve this problem in stages. S is closed iff $S' \subseteq S$. If S is a closed set, then, by part (a) and Lemma 1, $S' \subseteq S$. Now suppose $S' \subseteq S$. Let $\{\vec{s}_n\} \subseteq S$ and let $\lim_{n\to\infty} \vec{s}_n = \vec{y} \in E$. To prove that S is closed we need only show that $\vec{y} \in S$. There are two cases: If infinitely many of the $\vec{s}_n$ are distinct, then every neighborhood of $\vec{y}$ contains infinitely many points of S. Hence $\vec{y} \in S'$, but $S' \subseteq S$ so $\vec{y} \in S$ in this case. In the second case only a finite number of the $\vec{s}_n$ are distinct. But then there is an N such that $\vec{s}_n = \vec{s}_{n+1}$ for all $n \geq N$. Because $\lim \vec{s}_n = \vec{y}$, it follows that $\vec{y} \in S$.

$(S')' \subseteq S'$ and so S' is a closed set.

Let $\vec{y}_0 \in (S')'$. By (a) there is a sequence $\{\vec{y}_n\} \subseteq S'$ such that $\lim_{n\to\infty} \vec{y}_n = \vec{y}_0$ and the $\vec{y}_n$ are distinct. Given $\varepsilon > 0$ we must show that $O_\varepsilon(\vec{y}_0)$ contains infinitely many points of S; this shows that

$\vec{y}_0 \in S'$. Choose N such that $\left\|\vec{y}_0 - \vec{y}_N\right\| < \frac{\varepsilon}{2}$. Then the set of all $\vec{s} \in S$ such that $\left\|\vec{s} - \vec{y}_N\right\| < \frac{\varepsilon}{2}$ is infinite because $\vec{y}_N \in S'$. But for any such $\vec{s}$, $\left\|\vec{s} - \vec{y}_0\right\| \leq \left\|\vec{s} - \vec{y}_N\right\| + \left\|\vec{y}_N - \vec{y}_0\right\| < \varepsilon$.

(c) $\bar{S}$ is a closed set. If C is closed, $S \subseteq C$, then $\bar{S} \subseteq C$. Let $\vec{y}_0 \in (S \cup S')'$ and let $\{\vec{y}_n\} \subseteq S \cup S'$ be a sequence of distinct points that converges to $\vec{y}_0$. Infinitely many of the $\vec{y}_n$ are in S, or infinitely many of the $\vec{y}_n$ are in S'. In the first case $\vec{y}_0 \in S'$, and the second case $\vec{y}_0 \in (S')'$.

Now let C be a closed set that contains S. We need only show that $S' \subseteq C$; but this is immediate from (a) and Lemma 1.

(d) Let H be a linear subspace of E. Then $\bar{H}$ is a linear subspace. Given $\vec{y} \in \bar{H}$ and $\lambda \in \mathbb{k}$ we must show $\lambda\vec{y} \in \bar{H}$. Clearly, we may assume $\lambda \neq 0$. Now $\vec{y} \in \bar{H} = H \cup H'$. If $\vec{y} \in H$, we are done. If $\vec{y} \in H'$, we can find $\{\vec{h}_n\} \subseteq H$ such that $\lim_{n \to \infty} \vec{h}_n = \vec{y}$. Given $\varepsilon > 0$ there is an N such that $\left\|\vec{h}_n - \vec{y}\right\| < \frac{\varepsilon}{|\lambda|}$ for all $n \geq N$. But then $\left\|\lambda\vec{h}_n - \lambda\vec{y}\right\| = |\lambda| \left\|\vec{h}_n - \vec{y}\right\| < \varepsilon$ for all $n \geq N$. Thus $\lim \lambda\vec{h}_n = \lambda\vec{y}$ and because every $\lambda\vec{h}_n \in H$, $\lambda\vec{y} \in H' \subseteq \bar{H}$.

Now let $\vec{x}, \vec{y} \in \bar{H}$. We must show that $\vec{x} + \vec{y}$ is in $\bar{H}$. If $\vec{x}$ and $\vec{y}$ are in H there is nothing to do. If $\vec{x} \in H$, $\vec{y} \in H'$, we choose $\{\vec{y}_n\} \subseteq H$ such that $\lim \vec{y}_n = \vec{y}$. Then $\lim (\vec{x} + \vec{y}_n) = \vec{x} + \vec{y}$ and because $(\vec{x} + \vec{y}_n) \in H$ for all n, $(\vec{x} + \vec{y}) \in H'$. Finally, if $\vec{x}, \vec{y} \in H'$ choose $\{\vec{x}_n\} \subseteq H$ $\{\vec{y}_n\} \subseteq H$ such that $\lim \vec{x}_n = \vec{x}$ and $\lim \vec{y}_n = \vec{y}$. Clearly, $(\vec{x}_n + \vec{y}_n) \in H$ for all n, and $\lim (\vec{x}_n + \vec{y}_n) = \vec{x} + \vec{y}$.

3.3.1 (a) For any $\vec{u} \in V$ we have $\langle \vec{u}, \vec{0} \rangle = \langle \vec{u}, 2 \cdot \vec{0} \rangle = \overline{\langle 2 \cdot \vec{0}, \vec{u} \rangle} = 2\overline{\langle \vec{0}, \vec{u} \rangle} = 2\langle \vec{u}, \vec{0} \rangle$. If $\langle \vec{u}, \vec{0} \rangle \neq 0$, then we could divide and obtain $2 = 1$. Note: $\vec{v} + \vec{0} = \vec{v}$ all $\vec{v}$, so $\vec{0} + \vec{0} = \vec{0}$, so $2 \cdot \vec{0} = \vec{0}$.

(b)
$$\|\vec{u}+\vec{v}\|^2 + \|\vec{u}-\vec{v}\|^2 = \langle \vec{u}+\vec{v}, \vec{u}+\vec{v}\rangle + \langle \vec{u}-\vec{v}, \vec{u}-\vec{v}\rangle$$
$$= \langle \vec{u},\vec{u}\rangle + \langle \vec{u},\vec{v}\rangle + \langle \vec{v},\vec{u}\rangle + \langle \vec{v},\vec{v}\rangle$$
$$+ \langle \vec{u},\vec{u}\rangle + \langle -\vec{v},\vec{u}\rangle + \langle \vec{u},-\vec{v}\rangle$$
$$+ \langle -\vec{v},-\vec{v}\rangle$$
$$= \|\vec{u}\|^2 + \|\vec{v}\|^2 + \|\vec{u}\|^2 + \|\vec{v}\|^2$$
$$= 2\left(\|\vec{u}\|^2 + \|\vec{v}\|^2\right)$$

(c)
$$\|\vec{u}+\vec{v}\|^2 = \langle \vec{u}+\vec{v},\vec{u}+\vec{v}\rangle = \langle \vec{u},\vec{u}\rangle + \langle \vec{u},\vec{v}\rangle + \langle \vec{v},\vec{u}\rangle + \langle \vec{v},\vec{v}\rangle$$
$$= \|\vec{u}\|^2 + \|\vec{v}\|^2$$

because $\langle \vec{u},\vec{v}\rangle = \langle \vec{v},\vec{u}\rangle = 0$.

(d) $\|\vec{u}+\vec{v}\|^2 - \|\vec{u}-\vec{v}\|^2 = 2\langle \vec{u},\vec{v}\rangle + 2\langle \vec{v},\vec{u}\rangle$ as we see from (b).

Now $\|\vec{u}+i\vec{v}\|^2 = \langle \vec{u}+i\vec{v}, \vec{u}+i\vec{v}\rangle = \langle \vec{u},\vec{u}\rangle - i\langle \vec{u},\vec{v}\rangle + i\left[\langle \vec{v},\vec{u}\rangle - i\langle \vec{v},\vec{v}\rangle\right] = \|\vec{u}\|^2 - i\langle \vec{u},\vec{v}\rangle + i\langle \vec{v},\vec{u}\rangle + \|\vec{v}\|^2$
and so $i\|\vec{u}+i\vec{v}\|^2 = i\|\vec{u}\|^2 + \langle \vec{u},\vec{v}\rangle - \langle \vec{v},\vec{u}\rangle + i\|\vec{v}\|^2$
Also
$$\|\vec{u}-i\vec{v}\|^2 = \langle \vec{u}-i\vec{v}, \vec{u}-i\vec{v}\rangle = \langle \vec{u},\vec{u}\rangle + i\langle \vec{u},\vec{v}\rangle - i\left[\langle \vec{v},\vec{u}\rangle + i\langle \vec{v}.\vec{v}\rangle\right]$$
$$= \|\vec{u}\|^2 + i\langle \vec{u},\vec{v}\rangle - i\langle \vec{v},\vec{u}\rangle + \|\vec{v}\|^2 \text{ so } i\|\vec{u}-i\vec{v}\|^2$$
$$= i\|\vec{u}\|^2 - \langle \vec{u},\vec{v}\rangle + \langle \vec{v},\vec{u}\rangle + i\|\vec{v}\|^2$$

Thus $i\|\vec{u}+i\vec{v}\|^2 - i\|\vec{u}-i\vec{v}\|^2 = 2\langle \vec{u},\vec{v}\rangle - 2\langle \vec{v},\vec{u}\rangle$

and so
$$\frac{\|\vec{u}+\vec{v}\|^2 - \|\vec{u}-\vec{v}\|^2 + i\|\vec{u}+i\vec{v}\|^2 - i\|\vec{u}-i\vec{v}\|^2}{4} = \frac{4\langle \vec{u},\vec{v}\rangle}{4}$$
$$= \langle \vec{u},\vec{v}\rangle$$

3.3.3 $\{\vec{u}_n\}$ and $\{\vec{v}_n\}$ are sequences in V that converge to $\vec{u}_0$ and $\vec{v}_0$.

(a) For any $\vec{w} \in V$, $\lim \langle \vec{u}_n, \vec{w}\rangle = \langle \vec{u}_0, \vec{w}\rangle$. We write
$$|\langle \vec{u}_n, \vec{w}\rangle - \langle \vec{u}_0, \vec{w}\rangle| = |\langle \vec{u}_n - \vec{u}_0, \vec{w}\rangle| \le \|\vec{u}_n - \vec{u}_0\|\|\vec{w}\| \text{ by the}$$
C.S.B. inequality (Theorem 2).
Because $\vec{w}$ is fixed and $\lim \|\vec{u}_n - \vec{u}_0\| = 0$, we are done.

(b) $\lim \langle \vec{u}_n, \vec{v}_n\rangle = \langle \vec{v}_0, \vec{v}_0\rangle$. Here we write $|\langle \vec{u}_n, \vec{v}_n\rangle - \langle \vec{u}_0, \vec{v}_0\rangle| =$
$$|\langle \vec{u}_n, \vec{v}_n\rangle - \langle \vec{u}_n, \vec{v}_0\rangle + \langle \vec{u}_n, \vec{v}_0\rangle - \langle \vec{u}_0, \vec{v}_0\rangle| =$$
$$|\langle \vec{u}_n, \vec{v}_n - \vec{v}_0\rangle + \langle \vec{u}_n - \vec{u}_0, \vec{v}_0\rangle| \le |\langle \vec{u}_n, \vec{v}_n - \vec{v}_0\rangle|$$
$$+ |\langle \vec{u}_n - \vec{u}_0, \vec{v}_0\rangle| \le \|\vec{u}_n\|\|\vec{v}_n - \vec{v}_0\| + \|\vec{u}_n - \vec{u}_0\|\|\vec{v}_0\|.$$

Now $\|\vec{u}_n - \vec{u}_0\| \to 0$ as $n \to \infty$ so the last term tends to zero. Because $\{\vec{u}_n\}$ is convergent the set $\{\|\vec{u}_n\| \mid n = 1, 2, \ldots\}$ is bounded; i.e., there is an M such that $\|\vec{u}_n\| \le M$ all n; to see this just note that $\|\vec{u}_n\| = \|\vec{u}_n - \vec{u}_0 + \vec{u}_0\| \le \|\vec{u}_n - \vec{u}_0\| + \|\vec{u}_0\|$ and $\|\vec{u}_n - \vec{u}_0\| \to 0$. Because $\|\vec{v}_n - \vec{v}_0\| \to 0$ we see that the first term in our inequality tends to zero.

3.4.1 (a) S is an orthonormal subset of V. We claim S is maximal iff $\langle \vec{u}, \vec{v}\rangle = 0$ all $\vec{v} \in S$ implies $\vec{u} = \vec{0}$.

First suppose that S is maximal and let $\vec{u} \in V$, $\langle \vec{u}, \vec{v}\rangle = 0$ for all $\vec{v} \in S$. If $\vec{u} \ne \vec{0}$, then $S \subsetneq S \cup \left\{\frac{\vec{u}}{\|\vec{u}\|}\right\}$, and this last set is orthonormal. This contradicts the maximality of S.

Now suppose that the orthonormal set S has the property stated. If $S \subseteq S'$, where S' is orthonormal, we must show that $S = S'$. Suppose $\vec{u} \in S' - S$. Then $\|\vec{u}\| = 1$ because $\vec{u} \in S'$, but we also must have $\langle \vec{u}, \vec{v}\rangle = 0$ for all $\vec{v} \in S$. This implies $\vec{u} = \vec{0}$, which is a contradiction.

(b) Any $\vec{z} = \{z_j\}_{j=1}^\infty$ such that $\langle \vec{z}, \vec{e}_n \rangle = 0$ for all n must be the zero vector because $\langle \vec{z}, \vec{e}_n \rangle = z_n$ for each n. Thus, by (a) the set $\{\vec{e}_n\}$ is maximal.

(c) Let $f \in C(T)$ and suppose that $\langle f, \frac{e^{int}}{\sqrt{2\pi}} \rangle = \frac{1}{\sqrt{2\pi}} \int_{-\pi}^{\pi} f(t) e^{-int} dt = 0$ for every n. Then all the Fourier coefficients of f are zero. It follows that the partials sums of the Fourier series of f, and hence the Cesáro means of this series, are zero. Because these means converge uniformly to f (Theorem 1 of Section 7, Chapter 1) we see that f must be identically zero.

3.5.1 The linear span of the set $\{e^{int}\}$ in L^1 coincides with the set of trigonometric polynomials (see Chapter One, Section 7 just above Corollary 2). We want to show that this set is dense in $L^1[-\pi, \pi]$ for $\|\cdot\|_1$; i.e., we want to find, for any given $f \in L^1[-\pi, \pi]$, a sequence $\{\sigma_n\}$ that converges to f for the L^1-norm. Now $C(T)$ is dense in $L^1[-\pi, \pi]$ and so, for the given f, we can find $\{g_n\} \subseteq C(T)$ such that $\{g_n\}$ converges to f for the L^1-norm. For each fixed n we can find σ_n such that $\max_{-\pi < t \leq \pi} |g_n(t) - \sigma_n(t)| < \frac{1}{2\pi n}$ (Chapter One, Section 7, Corollary 2). Clearly, $\|g_n - \sigma_n\|_1 < \frac{1}{n}$ for each n. But then $\|f - \sigma_n\|_1 \leq \|f - g_n\|_1 + \|g_n - \sigma_n\|_1 \leq \|f - g_n\|_1 + \frac{1}{n}$, and we see that $\{\sigma_n\}$ converges to f for the L^1-norm.

3.5.2 For notational convenience let us set $\vec{v}_n = \frac{e^{int}}{\sqrt{2\pi}}$, for each n. By Theorem 5, $\{\vec{v}_n\}_{-\infty}^{\infty}$ is an orthonormal basis for $L^2(T)$. Consider the sequence $\{\sum_{-k}^{k} \beta_n \vec{v}_n\}_{k=0}^{\infty} \subseteq L^2(T)$. Because $\sum_{-\infty}^{\infty} |\beta_n|^2$ is convergent, this sequence is Cauchy (see the proof of Corollary 2 to Theorem 1 in Section 4). Thus $\sum_{-\infty}^{\infty} \beta_n \vec{v}_n$ converges to an element $f \in L^2(T)$. But then, by the discussion at the very beginning of Section 4, we must have $\langle f, \vec{v}_n \rangle = \langle f, \frac{e^{int}}{\sqrt{2\pi}} \rangle = \beta_n$ for every n.

Chapter Four

4.1.2 (a) $(a * b)(n) = \sum_m a(n-m) b(m) = \sum_k a(k) b(n-k) = (b * a)(n)$, where we have set $k = n - m$ (remember n is fixed).

(b) $\lambda(a * b) = \lambda\{(a * b)(n)\} = \lambda\{\sum_m a(n-m)b(m)\} = \{\sum \lambda a(n-m) b(m)\} = \{\sum a(n-m) \cdot \lambda b(m)\}$, and the first of these is $\lambda a * b$, whereas the second is $a * \lambda b$.

(c) $[a * (b * c)](n) = \sum_m a(n-m)(b * c)(m) = \sum_m a(n-m)(\sum_k b(m-k) c(k)) = \sum_k c(k) (\sum_m a(n-m) b(m-k))$ if we let $l = m - k$ this becomes

$\sum_k c(k) \left(\sum_l a[n-(l+k)] b(l)\right) =$
$\sum_k c(k) \left(\sum_l a[(n-k)-l] b(l)\right) = \sum_k (a*b)(n-k) c(k) =$
$[(a*b)*c](n)$.

(d) $[a*(b+c)](n) = \sum_m a(n-m)(b+c)(m) =$
$\sum_m a(n-m)\{b(m)+c(m)\} = \sum_m a(n-m)b(m) +$
$\sum_m a(n-m)c(m) = (a*b)(n) + (a*c)(n)$

(e) Suppose $a*c = c*a = e$ and $a*b = b*a = e$. Then $b = b*c = b*(a*c) = (b*a)*c = (e)*c = c$.

4.1.3 $\|a*b_n - a*b_0\|_1 = \|a*(b_n-b_0)\|_1 \leq \|a\|_1 \|b_n-b_0\|_1$. Because $\lim_{n\to\infty} \|b_n-b_0\|_1 = 0$, we see that $\lim_n a*b_n = a*b_0$ for every fixed $a \in l^1(Z)$.

If we define $T_a(b) = (a*b)$ for all $b \in l^1$, where $a \in l^1$ is fixed, then T_a is linear by 4.1.2 parts (b) and (d). This map is also continuous because $\|T_a(b)\|_1 = \|a*b\|_1 \leq \|a\|_1 \|b\|_1$ for all $b \in l^1$.

4.1.5 (a) $g(1) = 1$, $g(n) = 0$ all $n \neq 1$. Then $g*g(n) = \sum_m g(n-m)g(m)$, and this is zero unless $m = 1$ and $n - m = 1$. Thus $(g*g)(n) = 1$ when $n = 2$ and is zero when $n \neq 2$. Suppose $g^k(n) = 1$ when $n = k$ and is zero when $n \neq k$. Then $g^{k+1}(n) = g*g^k(n) = \sum_m g(n-m)g^k(m)$ is zero unless $m = k$ and $n - m = 1$. Thus $g^{k+1}(k+1) = 1$, and $g^{k+1}(n) = 0$ if $n \neq k+1$.

(b) $g^{-1}(n) = 1$ when $n = -1$, $g^{-1}(n) = 0$ when $n \neq -1$. Then $g*g^{-1}(n) = \sum_m g(n-m)g^{-1}(m)$, and this is zero unless $m = -1$ and $n - m = 1$. So $g*g^{-1}(0) = 1$, and $g*g^{-1}(n) = 0$ for $n \neq 0$; i.e., $g*g^{-1} = e$. Now $g^{-1}*g(n) = \sum g^{-1}(n-m)g(m)$ is zero unless $m = 1$ and $n - m = -1$. So $g^{-1}*g = e$ and we have shown that g^{-1} is the inverse of g.

(c) $(g^{-1})^2(n) = (g^{-1}*g^{-1})(n) = \sum g^{-1}(n-m)g^{-1}(m)$. This is zero unless $m = -1$ and $n - m = -1$. So $(g^{-1})^2(n) = 1$ when $n = -2$ and is zero when $n \neq -2$. As in (a) we see that $(g^{-1})^k(n) = 1$ when $n = -k$ and is zero when $n \neq -k$.

(d) $a \in l^1$ so $a = \{a(n)\}$. Also $\sum_{-\infty}^{\infty} a(n)g^n = \ldots + a(-1)g^{-1} + a(0)g^0 + a(1)g^1 + \ldots = \{a(n)\}_{-\infty}^{\infty}$, and this is just a.

4.2.1 We have $a \in l^1(z)$ fixed, $I(a) = \{a*b \mid b \in l^1\}$.

(a) Because $a*e = a$, $a \in I(a)$. Clearly, $a*b + a*c = a*(b+c)$, and $\lambda(a*b) = a*(\lambda a)$ by 4.1.2 (b) and (d). Thus $I(a)$ is a linear subspace of l^1. To show that $I(a)$ is an ideal, we must show that when $c \in I(a)$ and $d \in l^1$, $d*c \in I(a)$. Now $c \in I(a)$ and so $c = a*b$ for some $b \in l^1$. Also $d*c = d*(a*b) = (d*a)*b = (a*d)*b = a*(d*b)$ by 4.1.2 (c) and (a).

(b) We want to show $I(a)$ is a proper iff a is **not** invertible. First suppose that a is invertible. Then there is an $a^{-1} \in l^1$ such that $a*a^{-1} = a^{-1}*a = e$. Hence $e \in I(a)$. But then $b = b*e \in I(a)$ for every $b \in l^1$ showing that $I(a)$ is not proper. Thus when $I(a)$ is proper, a cannot be invertible. Now suppose that a is noninvertible. Then $e \notin I(a)$ for, otherwise, $e = a*b = b*a$ for some $b \in l^1$, which would mean a is invertible. Thus $I(a)$ must be proper.

(c) As discussed just before Lemma 1 we need only show that the closure of an ideal is an ideal. Let I be an ideal in l^1 and let $\bar{I}$ be the closure of I. Then I, and hence $\bar{I}$ (by 3.2.2 (f)), is a linear subspace of l^1. Let $a \in \bar{I}$ and let $b \in l^1$. We must show that $b*a$ is in I. If $a \in I$, this is immediate. If $a \in \bar{I} \setminus I$, then there is a sequence $\{a_n\} \subseteq I$ such that $\lim_n a_n = a$ for the l^1 norm. But then $b*a_n \in I$ for all n, and $\lim_n b*a_n = b*a$ for the l^1-norm (4.1.3). Thus $b*a \in \bar{I}$ and we are done.

4.3.4 Suppose that for some $e \in L^1(\mathbb{R})$ we have $e*f = f$ for every $f \in L^1(\mathbb{R})$. Then $e*e = e$. Hence $\hat{e}(x) \cdot \hat{e}(x) = \hat{e}(x)$ or $\hat{e}(x)[\hat{e}(x) - 1] = 0$ for all $x \in \hat{\mathbb{R}}$. Now $\hat{e}(x)$ is a continuous function (Lemma 1), so either $\hat{e}(x) = o$ for all x or $\hat{e}(x) = 1$ for all x. The first of these would mean that e is the zero element of $L^1(\mathbb{R})$ and in that case $e*f = 0 \neq f$. The second possibility can be ruled out because $\lim_{|x| \to \infty} \hat{e}(x) = 0$ by Lemma 2.

4.4.1 We refer to the statement of Lemma 2. Here M is a multiplicative group. Because $M \neq \emptyset$ there is an $x \in M$. But then $x \cdot x^{-1} = 1 \in M$ and we have proved (i). To prove (ii) we note that for any $x \in M$, $1 \cdot x^{-1}$ is in M. When x, y are in M, x and y^{-1} are in M by (ii). Hence $x \cdot \left(y^{-1}\right)^{-1} = x \cdot y$ is in M. It is now clear that $x^n \in M$ for every $n \in \mathbb{Z}$.

4.4.2 $f : M \to A$ where M is a multiplicative group and A is an additive group. Furthermore $f(x \cdot y) = f(x) + f(y)$ all x, y in M.

(a) For any $x \in M$, $f(x) = f(x \cdot 1) = f(x) + f(1)$ showing that $f(1) = 0$ and hence 0 is in the range of f. Also $f(1) = f(x \cdot x^{-1}) = f(x) + f(x^{-1})$, so $f(x^{-1}) = -f(x)$.

Now let a, b be in the range of f. We must show that $a - b$ is in the range of f. There are x, y in M such that $f(x) = a$ and $f(y) = b$. Hence $f(x \cdot y^{-1}) = f(x) + f(y^{-1}) = a + (-b)$ and we are done.

(b) $\{x \in M \mid f(x) = 0\} \neq \emptyset$ because it contains 1 (see (a)). If x, y are in this set, then $f(x \cdot y^{-1}) = f(x) + f(y^{-1}) = f(x) + [-f(y)]$. But both $f(x)$ and $f(y)$ are zero, so $x \cdot y^{-1}$ is in our set.

(c) $U_4 = \{\pm 1, \pm i\}$. Note that $i^{-1} = \frac{1}{i} = \frac{-i^2}{i} = -i$ and $(-i)^{-1} = \frac{-i^2}{-i} = i$.

(d) $f: Z \to U_4$, $f(n) = i^n$ all n. Then $f(m+n) = i^{m+n} = i^m \cdot i^n = f(m) \cdot f(n)$. For positive n, $i^n = i^{4q+r} = (i^4)^q \cdot i^r = i^r$ because $i^4 = 1$. Here $0 \le r < 4$ so i^n is i^0, i^1, i^2 or i^3. All of these, $i^0 = 1, i, i^2 = -1$, and $i^3 = -i$, are in U_4. So the range is U_4, because $i^{-n} = \frac{1}{i^n}$, and this is in U_4 for each positive n. When $n = 0$, $f(n) = 1 \in U_4$.
$\{n \in Z \mid f(n) = 1\} = \{n \in Z \mid i^n = 1\} = \{n \in Z \mid i^{4q+r} = 1\} = \{n \mid n = 4q + r, \text{ and } r = 0\} = $ all integer multiples of four.

(e) $g: Z \to U_4$, $g(n) = (-1)^n$. Here the range of g is $\{+1, -1\}$ and $\{n \in Z \mid g(n) = 1\} = $ all even integers.

4.4.3 Here n is a fixed positive integer and $U_4 = \{z \in \mathbb{C} \mid z^n = 1\}$.

(a) If $x, y \in U_n$, then $x^n = 1$ and $y^n = 1$. Also $(y^{-1})^n = \frac{1}{y^n} = 1$. Thus $(x \cdot y^{-1})^n = x^n \cdot (y^{-1})^n = 1$ and hence $x \cdot y^{-1}$ is in U_n. If $Z^n = 1$, then $1 = |Z^n| = |Z|^n$ by 0.4.4 (a). Thus $|Z| = 1$ because the only positive real number whose nth power is 1, is 1.

(b) If $p(x) = 0$ has a root r of multiplicity $k > 1$, then $p(x) = (x-r)^r q(x)$. But then $p'(x) = (x-r)^k q'(x) + k(x-r)^{k-1} q(x)$ by the product rule. Because $k > 1$, $k - 1 > 0$ and we see that $p'(r) = 0$.

Now set $p(x) = x^n - 1$. Because $p'(x) = nx^{n-1}$ its only solutions are $x = 0$ (with multiplicity $n-1$). Because zero is clearly not a root of $p(x)$, $p(x)$ and $p'(x)$ have no common roots. Thus all roots of $p(x)$ are simple.

4.4.5 (a) Clearly, $\{i^n \mid n \in Z\} = \{i^0, i^1, i^2, i^3\} = U_4$, so i is a generator. So also is $-i$.

(b) For n fixed, $e^{i\frac{2\pi}{n}} = \cos\frac{2\pi}{n} + i\sin\frac{2\pi}{n}$ is clearly in U_n; for $\left(e^{i\frac{2\pi}{n}}\right)^n = e^{2\pi i} = \cos 2\pi + i\sin 2\pi = 1$. But the n numbers $\left(e^{i\frac{2\pi}{n}}\right)^k$, $0 \le k < n$ are in U_n, and they are distinct. So we have a generator of U_n.

(c) The number 1 is a generator of the additive group Z.

4.5.1 M is a multiplicative group; N is a subgroup of M.

(a) For $s, t \in M$, $s \equiv t \bmod N$ is to mean $st^{-1} \in N$. Because N is a multiplicative group, $1 \in N$ (Section 4, Lemma 2). Thus for any $s \in M$, $s \cdot s^{-1} \in N$ and so $s \equiv s \bmod N$. Suppose $s \equiv t \bmod N$. Then $s \cdot t^{-1} \in N$. But then $(s \cdot t^{-1})^{-1} \in N$ (Section 4, Lemma 2) and so $t \cdot s^{-1} \in N$. It follows that $t \equiv s \bmod N$. Finally, let $s \equiv t \bmod N$, $t \equiv u \bmod N$. Then st^{-1} and tu^{-1} are in N. But then $(st^{-1})(tu^{-1}) = su^{-1}$ is in N, and we see that $s \equiv u \bmod N$ (again we used Lemma 2 of Section 4).

(b) Let $s' \in [s]$, and $t' \in [t]$. Then $s's^{-1} \in N$ and $t't^{-1} \in N$. Hence $\left(s's^{-1}\right)\left(t't^{-1}\right) = (s't')(st)^{-1} \in N$ showing $s't' \equiv st \bmod N$.

(c) For any $s \in M$, $[s] * [1] = [s \cdot 1] = [s] = [1 \cdot s] = [1] * [s]$.

(d) For any $s \in M$, $s^{-1} \in M$ (Lemma 2, Section 4 of Chapter 4). Hence $[s] * [s^{-1}] = [s^{-1}] * [s] = [ss^{-1}] = [s^{-1}s] = [1]$.

(e) For $z_1, z_2 \in \mathbb{C}\setminus\{0\}$, $z_1 \equiv z_2 \bmod N$ means $z_1 \cdot z_2^{-1} \in N$, which means $\frac{z_1}{z_2} \in N$. Now $z_1 = r_1 e^{i\theta_1}$, $z_2 = r_2 e^{i\theta_2}$, and $N = T$ (the unit circle). So $\frac{z_1}{z_2} = \frac{r_1}{r_2} e^{i(\theta_1-\theta_2)}$, and this will be in T iff $\frac{r_1}{r_2} = 1$. So our equivalence classes are concentric circles centered at the origin.

4.5.4 C is a subgroup of the additive group A. $\varphi : A \to \frac{A}{C}$ is defined by setting $\varphi(a) = [a]$ all $a \in A$. Then $\varphi(a+b) = [a+b] = [a] + [b] = \varphi(a) + \varphi(b)$ for all $a, b \in A$. Also $\{a \in A \mid \varphi(a) = [0]\} = \{a \in A \mid a \equiv 0 \bmod C\} = \{a \in A \mid a \in C\} = C$.

4.6.1 (a) Let $r = \sum_{j=0}^{l-1} \omega^j$ and so $\omega r = \sum_{j=0}^{l-1} \omega^{j+1}$. Now we may list the summands as follows: $r = \omega^0 + \omega^1 + \omega^2 + \omega^3 + \ldots + \omega^{l-2} + \omega^{l-1}$ and $\omega r = \omega^1 + \omega^2 + \omega^3 + \ldots + \omega^{l-2} + \omega^{l-1} + \omega^l$.

Observe that they differ only in that ω^0 is in the sum defining r and ω^l is in the sum-defining ωr. But $\omega^0 = 1 = \omega^l$ so they are equal.

(b) $\omega^n - 1 = (\omega - 1)\left(\omega^{n-1} + \omega^{n-2} + \ldots + \omega + 1\right)$. Now if $\omega^n = 1$, then at least one of the factors on the right must be zero. But $\omega \neq 1$; hence the second factor must be zero.

4.6.2 We have $\omega \in U_l$. (a) $\omega^{-1} = \frac{1}{\omega} = \frac{\omega^l}{\omega} = \omega^{l-1}$; (b) $\omega \cdot \bar{\omega} = |\omega|^2 = 1$ and so $\bar{\omega} = \frac{1}{\omega}$; (c) the order of each $\omega \in U_l$ is a division of l (Definition 5, Lemma 4). When l is a prime every member of U_l, except $+1$, has order l. Thus every such member of U_l. is primitive l^{th} root of unity.

4.8.2 If $\bar{a}, \bar{b}$ are in Z_l and $\bar{a} \cdot \bar{b} = \bar{0}$, then if $\bar{a}$ has an inverse $(\bar{a})^{-1}$ we would have $(\bar{a})^{-1}(\bar{a} \cdot \bar{b}) = (\bar{a})^{-1} \cdot \bar{0} = \bar{0}$. But $(\bar{a})^{-1} \cdot (\bar{a} \cdot \bar{b}) = \left[(\bar{a})^{-1} \cdot \bar{a}\right] \cdot \bar{b} = \bar{1} \cdot \bar{b}$ giving us $\bar{b} = \bar{0}$. Similarly, $\bar{b}$ cannot have inverse.

4.8.3 Suppose $ka \equiv kb \bmod n$ and $(k,n) = 1$. Then $\bar{k} \cdot \bar{a} = \bar{k} \cdot \bar{b}$ in Z_n because $(k,n) = 1$, $\bar{k}$ has an inverse $(\bar{k})^{-1} \in Z_n$ (see Theorem 3). But then $(\bar{k})^{-1} \cdot (\bar{k} \cdot \bar{a}) = (\bar{k})^{-1} \cdot (\bar{k} \cdot \bar{b})$, giving us $\bar{a} = \bar{b}$ or, equivalently, $a \equiv b \bmod n$.

4.9.2 We have $n = p \cdot q$ where p is the smallest prime factor of n. Here q is a prime or, if not, q has a prime factor q_1. Clearly, $p \leq q_1$ because q_1 is a prime factor of n. But then $p^2 \leq pq_1 \leq pq = n$.

4.9.5 If n is composite, then $n = pq$ where $1 < p < n$, $1 < q < n$. Thus $2^n - 1 = 2^{pq} - 1 = (2^p)^q - 1 = (2^p - 1)\left[(2^p)^{q-1} + (2^p)^{q-2} + \ldots + 1\right]$. Because $2^p - 1 > 0$ and the second factor is also positive, $2^n - 1$ is composite.

Index

A
Abel kernel, 126
Abelian group; see Group, abelian
Abel's test, 47(5e)
Absolute value, 10(3), 13, 14(4)
Abstract algebra, 165
Abundant number, 156
Additive group; see Group, addititive
Additive identity, 7, 13, 179
Additive inverse, 7, 13
Adherent point, 89(2)
Analysis, 15–21
Analytic geometry, 14
Angle, 13, 14, 90, 92
Approximation theorem (Weierstrass), 55, 65, 66
Argand, R. 13, 21
Associative law, 166, 179
Axiom of choice, 5, 6, 11(4), 121, 163
Axiom of continuity, 9
Axiomatic set theory, 7(4c)
Axioms for a field, 175
Axioms for a group, 165
Axioms for an algebra, 173
Axioms for a ring, 172
Axioms for a vector space, 179
Automorphism, 172(2)

B
Ball, closed, 83
Ball, open, 83, 84(3b)
Ball, radius epsilon, 83
Ball, unit, 84
Banach space, 89, 104
Basis, orthonormal, 101, 106, 108, 109
Belongs to, 1
Bernoulli, 23, 31, 55
Bernstein polynomials, 52(9)
Bessel, 13, 21
Bessel's inequality
  (strong form), 97
  (weak form), 92
Binary operation, 165
Blahut, E., 177
Boas, R.P., 21
Bombelli, 12, 21
Bose and Kim, 164
Bound, greatest lower, 10(2d)
Bound, least upper, 8, 9, 10(2c)
Bound, lower, 8, 10(2d)
Bound, upper, 8
Boundary data, 25, 51, 52, 54
Bounded function, 88
Bounded linear map, 86
Bounded sequence, 21(7d), 42(2), 46(4, 5)
Bounded set, 15, 21(7d)
Bruce, J.W., 163
Brunelleschi, 11
Bunyakowski, (C.S.B. inequality), 92
Butzer, P.L. and Stark, E.L., 55

C
Cantor set, 11(7)
Cardano, G., 12, 21
Cartesian product, 3, 4, 13
Cauchy, A.L., 18

Cauchy (C. S.B. inequality), 92
Cauchy-Riemann equations, 18, 20(4b)
Cauchy sequence, 19, 20(1, 6a), 21(7), 88, 181
Cesàro, E., 43
Cesàro convergent, 43, 46(1, 2, 4)
Cesàro mean, 43, 44, 47
Cesàro summable, 43
Character of the additive group of integers, $(\mathbb{Z}, +)$, 137
Character of the additive of integers modulo n, $(\mathbb{Z}_n, +)$, 137
Character of the additive group of real numbers, $(\mathbb{R}, +)$, 143(7)
Characteristic of a field, 175
Circle, 24
Circle, unit (T), 25, 27
Classical harmonic analysis, 23–55
Closed ball, 83, 84(3b)
Closed set, 15, 83, 84(3), 85, 90(4b)
Closed under a binary operation, 165
Closure of an ideal, 121
Closure of a set, 89(2)
Closure of a subspace, 89(2d)
Codomain, 5
Collection, 1
Commutative group; see Group, Abelian
Commutative law, 179
Commutative ring, 172
Compact set, 16
Compact support, 107(3b), 108
Comparable elements, 6(3b)
Complement, orthogonal, 109, 111, 112(1)
Complement, set-theoretic, 3(4), 15
Complete, inner product space, 97, 105
Complete normed space, 89
Complete orthonormal set; see Maximal orthonormal set
Completeness property of $\mathbb{R}$, 9
Completion, 104, 105, Appendix B, 181–186

Complex conjugate, 14(1)
Complex Fourier coefficients, 41, 51
Complex Fourier series, 41, 51
Complex numbers ($\mathbb{C}$), 11–14
Composite number, 153, 155
Composition of functions, 6(2)
$\mathbb{C}^n$(complex n-space), 79,
$\mathbb{C}^n$(complex n-space), norm of, 82(iii)
Congruence modulo a subgroup, 131, 167
Contained in, 2
Continuous function, 16, 17, 19, 20(2, 3b, 5), 26, 27, 28(3), 34
Continuous linear map, 86
Convergent sequence, 19, 20(1), 21(7a, c), 85
Convergent series, 33, 37(3, 4)
Convolution, 114, 118(2–5), 126, 128(3), 135
$C_r(T)$, Continuous (real-valued) functions on the unit circle, 39, 41, 42, 47, 49, 50, 51, 52
C(T), continuous functions on the unit circle, 25–28, 51(remark 1), 80, 89
$C_0(\mathbb{R})$, continuous functions on $\mathbb{R}$ with compact support, 107 (3b)
Cubic equation, 145
Cyclic group; see Group, cyclic

D

D'alembert, 23, 31, 55
Danish Academy of Sciences, 13
Dantzig, T., 21
Debnath, L. and Mikusinski, P., 112
Deficient number, 156
del Ferro, 12, 21
De Morgan, 3(4a), 7(5)
Dense set, 104, 105
Derivative, 16, 17, 35, 36, 37, 42(3)
Derivative, one-sided, 68, 69, 73, 74
DeVito, C.L., 112
Dickson, L.E., 164

Differentiable, 16, 17, 20(2, 3c), 35, 42(3), 126
Differential equation, ordinary, 29, 30, 31, 31, 127
Differential equation, partial, 18, 25, 30, 32
Dilation, 108
Direction, 90
Dirichlet kernel, 70
Dirichlet problem (DP), 24, 25, 29, 47, 52
Disjoint sets, 2
Disjoint sets, pairwise, 5
Distance, 14
Divergent, 19
Divisor (factor) of a natural number, 154
Divisor of an integer, 150
Divisor of zero, 153
Domain of a function, 5, 6(2)
Domain of a relation, 4

E

Edwards, H.M., 163
Equal sets, 2
Equivalence class, 6(1), 103, 132, 134(1), 167
Equivalence relation, 4, 6(1, 3), 132, 134(1), 167
Equation, Laplace's, 18, 20(4a), 24
Equation, ordinary differential; see Differential equation, ordinary
Equation, partial differential; see Differential equation, partial
Equation polynomial, 12, 143
Equations, Cauchy-Riemann; see Cauchy-Riemann equations
Euclid, 156
Euclidean norm, 81(i), 82(iii)
Euler, L., 13, 21, 23, 31, 55, 159(3c, 6), 161
Euler formulas, 39, 40, 41, 57
Euler's phi function, 154, 159, 164
Euler's totient function; see Euler's phi function

Even function, 60, 61(2, 3), 62, 64(2)

F

Factor; see Divisor
Family of sets, 5, 6, 7, 15, 16
Fejér, L., 47
Fejér's kernel, 46(3), 70
Fejér's theorem, 47, 62, 102(1c), 106
Fermat, P., 159, 161
Fermi paradox, 47, 55
Ferarri, L., 12
Field, 175
Finite dimensional, 79–82, 84(4)
Fontana, N. (Ta'taglia), 12, 21
Fourier, J-B.J., 57, 77
Fourier coefficients, 41, 42, 42(2, 3), 51, 58, 61, 68
Fourier cosine series, 62, 63
Fourier series, 41, 47, 49, 51, 58, 61, 70, 73
Fourier sine series, 62, 63
Fourier transform, 85
Fourier transform on $\mathbb{R}$, 123
Fourier transform on Z, 113
Fourier transform on Z_n, 139
Frederick the great, 163
Fredholm, 99
Function, 4, 5, 15(6), 23, 57
Function space, 113
Function, symmetric, 144
Functional, linear, 87
Functional, multiplicative, 120
Functions, addition of, 80, 113
Functions, multiplication of, 80, 115
Functions, multiplication by scalars, 80, 113
Functions, L.P., 103

G

Galois field, 176
Gauss, K.F., 13
Generator of a cyclic group; see Group, cyclic
Geophysicists, 108
Gerver, J., 37, 55

Gibbs phenomenon, 76(2)
Goldberg, R.R., 163
Greatest common divisor, 150, 151
Greatest lower bound, 10(2d), 27
Group, Abelian, 167
Group, abstract, 165
Group, additive, 128
Group, cyclic, 131(5)
Group, multiplicative, 129

H

Haar wavelet, 109
Hardy, G.H., 37
Harmonic analysis, 23
Harmonic function, 24, 37(1, 2)
Harriot, T., 12, 21
Hausdorff maximal principle, 121
Herstein, J.N., 177
Hewitt, E. and Ross, K.A., 163
Hilbert, D., 99
Hilbert space, 97, 105
Homomorphism, 169
Hungarians, 47

I

Ideal, proper, 121, 123(1), 173
Ideal, maximal, 121
Identity element, 166, 172
Identity operator, 87
Imaginary part of a complex number, 15(6)
Independent set; see Linearly independent set
Inequality, Bessel's, (strong) 97, (weak) 92
Inequality, Cauchy-Schwarz-Bunyakowski, 92
Inequality, triangle, 10(3a), 14(4b), 81
Infinite dimensional spaces, 80, 84(4)
Inner product, 79, 91
Integers (Z), 9
Integral equations, 99
Integral, Lebesgue, 41, 103, 124
Integral, Riemann, 17, 20(3b), 35, 57
Integrable, 57

Intersection, 2
Invariant, 144, 145
Inverse, additive, 7, 13, 114, 125
Inverse element, 116, 119, 166
Inverse function, 6(2b), 51
Inverse, multiplicative, 8, 13
Invertible, 116, 120
Isomorphic, 170
Isomorphism, 170

J

Jeffery, R.L., 55, 77

K

Katznelsen, Y., 163
Kelley, J., 163
Kernel, Dirichlet's, 70
Kernel, Fejér's, 44, 70
Kernel, Poisson's, 55(1)
Kernel (null space), Appendix A, 90(4a), 180
Kernel of a homomorphism, 170
Konopinski, E., 47
Kreyszig, E., 77

L

Lagrange, G., 143, 163
Lagrange's identity, 71
Lagrange's solution to the cubic, 145
Laplace's equation, 18, 24, 30
Laplace's equation in polar co-ordinates, 25
Lebesgue integral; see Integral, Lebesgue
Left-shift operator, 87
Limit, 16, 20(3), 41, 42
Linear combination, Appendix A, 107
Linear functional, 87
Linear mapping, Appendix A, 85
Linear operator, 87
Linear subspace, Appendix A, 89(2d), 90(4a)
Linearly independent set, 84(4), 89(3)
$l^1(Z)$ (l-one of Z), 82(iv), 84(4), 113, 116

Lomont, J.S., 177
L.P. function, 103
L.P. spaces, 103, 104
$l_{\mathbb{C}}^2(\mathbb{N})$ (complex l-2 of $\mathbb{N}$), 95(6)
$l_{\mathbb{R}}^2(\mathbb{N})$ (real l-2 of $\mathbb{N}$), 82(ii), 82(iii), 84(4), 93, 101
$l_{\mathbb{C}}^2(\mathbb{Z})$ (complex l-2 of $\mathbb{Z}$), 82
$l_{\mathbb{R}}^2(\mathbb{Z})$ (real l-2 of $\mathbb{Z}$), 82
$L^1(\mathbb{R})$ (L-1 of $\mathbb{R}$), 107(4), 123
$L^1(T)$ (L-1 of T), 104
$L^2(\mathbb{R})$ (L-2 of $\mathbb{R}$), 107(3c)
$L^2(T)$ (L-2 of T), 105, 106
Lucas-Lehmer test, 157

M

MacLaurin's series, 50
Matrix, 139, 141, 168, 172
Maximal ideal; see Ideal, maximal
Maximal Nest, 121
Maximal orthonormal set, 98, 101, 102(1)
Maximum, 17, 24
Mersenne numbers, 157, 163
Meyer, Y., 112
Minimum, 17, 24
Modulus, 13
Monomorphism, 170
Morphorism, 169
Multiplicity of a root, 131(3b)
Multiresolution analysis, 108, 109

N

Natural numbers ($\mathbb{N}$), 8
Neighborhood, 15, 83
Nest, 121
Norm, 81
Norm, Euclidean, 81(i), 82(iii)
Norm, from an inner product, 93
Norm, sup, 88
Normed space, 81
Normalized vector, 91
Normal subgroup; see Subgroup, normal
Null space (kernel), Appendix A, 180

O

Odd function, 60, 61(2, 3), 62, 64(2)
One-to-one function, 5
Onto function, 5, 7(4)
Open ball, 83, 84(3a)
Open interval, 11(6)
Open neighborhood, 15
Open set, 15, 83, 84(3)
Operator, left-shift, 87
Operator, right-shift, 87
Order of an nth root, 138
Orthogonal complement, 109, 111, 112(1)
Orthogonal set, 91
Orthogonal vectors, 91
Orthonormal basis, 101, 106
Orthonormal set, 91, 96(infinite), 98(maximal)

P

Pairwise disjoint sets, 5
Paradox, Fermi's, 47, 55
Paradox, Russell's, 7(4c)
Parallelogram identity, 94(b)
Parseval's relation, 99(c and d)
Partial differential equation; see Differential equation, partial
Partial function, 4, 5, 6(2)
Partial order relation, 6(3b)
Perfect numbers, 156
Periodic function, 25, 27, 28(1–7), 29
Permutation, 145, 149(2, 3), 168
P(f), set of periods of a function f, 25, 26
Piecewise continuous function, 65, 66, 68, 69, 70(4), 73, 74
Poisson's integral, 55(1)
Poisson's kernel, 55(1)
Pointwise convergence, 49, 70, 73
Polar form of a complex number, 14
Polar identity, 94(1d)
Polynomial, 12, 50, 52(7), 65
Polynomial, trigonometric, 49, 50, 51(1, 2)
Preimage, 5, 90(4)

Prime numbers, 151, 154
Primitive root of unity, 138, 151
Privalov, I.I., 24
Product, Cartesian; see Cartesian product
Product, inner, 79, 91
Pythagorean relation, 94(1c)

Q

Q (rational numbers), 9, 11(5)
Quadratic equation, 144
Quantum mechanics, 99, 112

R

Rational numbers (Q), 9
Real numbers ($\mathbb{R}$), 7
Real part of a complex number, 15(6)
Relation, 4
Relation, equivalence; see Equivalence relation
Relation, partial order; see Partial order relation
Relation, total order; see Total order relation
Relatively prime integers, 151
Resolvent set, 123(2)
Riemann, B., 36, 41, 68
Riemann integral; see Integral, Riemann
Right-shift operator, 87
Ring, 172
Ring, commutative, 172
$\mathbb{R}^n$(real n-space), 79
$\mathbb{R}^n$(real n-space) norm of, 81(i)
Roots of an equation, 12
Roots of unity, 131(3), 138
Russell paradox; see Paradox, Russell's
Ryan, R.D., 112

S

Scalar multiplication, 179
Schroeder, M.R., 164
Schwarz (C.S.B. inequality), 92
Sectionally continuous function; see Piecewise continuous function
Sequence, 18
Sequence, bounded, 21(7d), 42(2)
Sequence, Cauchy, 19, 20(5, 6), 21(7), 88
Sequence, convergent, 19, 49, 85
Sequence, divergent, 19, 46(1)
Sequence, of functions, 32, 33, 39, 49
Series, 23, 33
Series, convergent, 33, 39
Series, Fourier, 41, 58, 61, 70
Series, of functions, 33, 39
Series, of numbers, 43, 46(5)
Series, power, 37(3, 4, 5)
Series, summable, 43
Set, 1, 2, 3
SETI, 47
Signal processing, 108
Silard, L., 47
Simmons, G.F., 163
Singular element, 120
Space, Banach, 88, 89
Space, Hilbert, 97
Space, inner product, 91
Space, normed, 81
Space, of functions, 80
Space, vector, Appendix A, 179
Span, linear, Appendix A, 180
Spectrum, 123(2)
Subgroup, 129, 166
Subgroup, normal, 171, 172(4)
Subset, 2
Subspace, Appendix A, 180
Sums, infinite, 31
Superposition, 107 (see also linear combination)
Support, 107(3b)
Symmetric function; see Function, symmetric

T

Ta'taglia (N. Fontana), 12, 21
Tauberian theorems, 46(4)

Teller, E., 47
Total order relation, 6(3b), 8
Totient function; see Euler's totient function
Transform, Fourier; see Fourier transform
Translation, 108
Triangle inequality, 10(3a), 14(4b), 81
Trigonometric polynomial; see Polynomial, trigonometric
Trigonometric series, 23

U

U_n, 131(3), 138
Uniformly continuous function, 17, 20(6), 28(7), 65, 69, 70(4)
Uniformly convergent sequence, 32, 33, 36, 37, 37(1), 39, 47, 51(1, 9)
Uniformly covergent series, 33, 39
Union, 2, 5
Unit circle (T), 27
Unit, (ring with unit), 172

V

Vandermonde, 163
Vector, Appendix A, 179

Vector, normalized, 91
Vector, unit, 91
Vector space, Appendix A, 179
Vector space of functions, 80
Vieta, 144, 146

W

Wavelet, 108
Webb, S., 55
Weierstrass approximation theorem, 55, 65, 66
Weierstrass, K., 36
Weierstrass M-test, 33, 36, 37
Wessel, 13, 21
Wiener, N., 115, 123, 163
Witmer, T.R., 21
Wojtaszczyk, P., 112

Y

York, H., 47

Z

Z (the integers), 9
Zero, proper divisor of, 153
Zero, vector, 179
Z_n (the integers modulo n), 133